THE SMART GUIDE TO

Patents

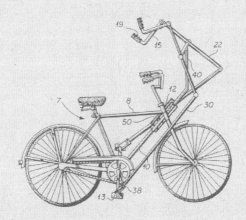

BY AARON G. FILLER

The Smart Guide To Patents

Published by

Smart Guide Publications, Inc.
2517 Deer Chase Drive
Norman, OK 73071
www.smartguidepublications.com

For information, address: Smart Guide Publications, Inc. 2517 Deer Creek Drive, Norman, OK 73071

SMART GUIDE and Design are registered trademarks licensed to Smart Guide Publications, Inc.

International Standard Book Number: 978-0-9834421-0-3

Library of Congress Catalog Card Number:
11 12 13 14 15 10 9 8 7 6 5 4 3 2 1

Printed in the United States of America

Cover design: Lorna Llewellyn
Copy Editor: Ruth Strother
Back cover design: Joel Friedlander, Eric Gelb, Deon Seifert
Back cover copy: Eric Gelb, Deon Seifert
Illustrations: James Balkovek
Production: Zoë Lonergan
Indexer: Cory Emberson
V.P./Business Manager: Cathy Barker

ACKNOWLEDGMENTS

My father was a successful inventor who passed along to me some of the books he read to get started. He was a civilian physicist with the US Navy who first trained for the invasion of Japan in World War II as a young sailor. He then filed a series of patents in the field of applied shock wave physics involving explosives for industrial, military and aerospace applications. The later redirection of this work to treat kidney stones in his lithotripsy patent was a classic innovative re-invention.

When I had just filed my first patent provisional application in the field of nerve medications, he gave me a key piece of advice. "Your invention is like your own child—no one cares about its success the way you do. It's your job to get it prepared so it can go out into the world and stand on its own."

To succeed in inventing and writing, I've been fortunate to have the support of my wife Lise and our two kids Rachel and Wyatt. My business backers including Hermann Hauser and Grant Hieshima, my co-inventors—prominently—Andrew Lever and Franklyn Howe, and my business partner Shirlee Jackson have also been critical in turning dreams into reality. I also thank my various publishers for their confidence and my literary agent Jodie Rhodes for her enthusiasm and support.

My education benefited from our CEOs—including Constance McKee and Tom Saylor, as well as from my Harvard professors—David Pilbeam, Irven DeVore, Fuzz Crompton, and Stephen Jay Gould. My courses at Concord Law School have been a great preparation as well. I've worked with gifted drafting patent agents—including Robin Waldren at Marks & Clerk, and attorneys Julian Cochbain at FB Dehn and Robert Perry of Gill Jennings & Every in the UK, as well as Dan Crouse and Steve Pollinger at Christenson O'Connor Johnson and Kindness in Seattle. I've also learned from patent litigators Eric Videlock of Pepper Hamilton, our great team at Russ August Kabat including Marc Fenster, Andrew Weiss, Alex Giza, and Fredricka Ung, as well as from strong opponents such as Greg LoCascio of Kirkland and Ellis and from demanding and precise judges such as Mariana Pfaelzer. Justin Strassburg has placed faith in my legal innovations that led to our breakthrough success in piercing the veil of sovereign immunity and forcing California into federal court to answer for patent infringement.

Archimedes famously stated, "Give me a place to stand and I can move the world." This is what my father, my family, my business associates, my legal associates, and my professors have done—they have truly given me a place to stand.

Aaron Filler
Santa Monica, California

TABLE OF CONTENTS

10 Inventor and Attorney Communication 99

11 Understanding the Language of Patent Claims 109

Disclaimer

Patent law is complex and ever-changing, and you are directed to retain your own patent attorney or agent and to rely exclusively on their advice before taking any action that affects or relates to any invention you have or may have. The information in this book is intended as an informational and educational guide to help the reader gain familiarity with the important terms and concepts that relate to patents. Although Smart Guide Publications and Aaron Filler have made every possible effort to assure that all of the information in this book is correct, accurate, and up to date, neither Smart Guide or the author, Aaron Filler, provide any warranty or guarantee regarding any of the information in the book or your reliance upon it. You should consult an attorney whom you retain before taking any action with any consequence in terms of intellectual property rights or financial impact. By purchasing this book and choosing to read, you agree to hold Smart Guide Publications and Aaron Filler harmless as to any damages, direct, consequential, or otherwise as they may impact you in consequence of or in relation to any information in this book.

PART ONE

Overview of Patents

FIG.1

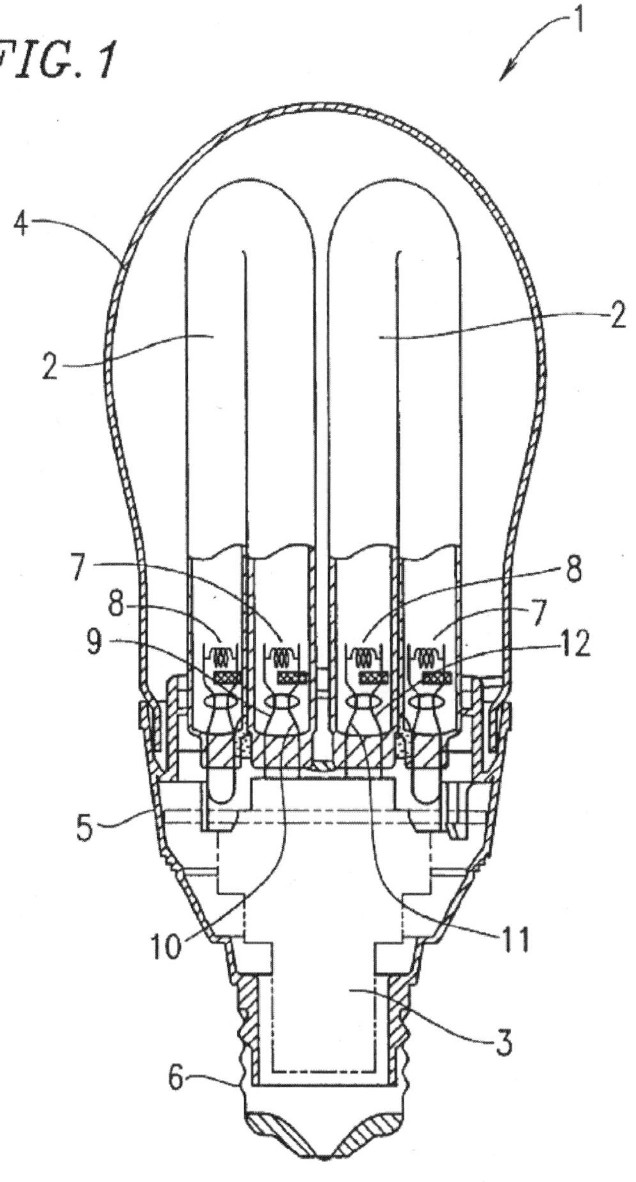

CHAPTER 1

What Exactly Is a Patent?

In This Chapter

➤ What a patent really is

➤ The role of the United States Patent Office

➤ How a person makes money from a patent

➤ Why the legal system gets involved in the process

Your First Questions

If you're reading this, you have probably just gone through the thrill of realizing that you may have thought of something that no one ever thought of before. If it still seems like a great original idea twenty-four hours later, then it's even more compelling.

Does your new idea make you an inventor? Is your idea patentable? Can it make you a fortune? No doubt these questions have occurred to you, and they have also led you to read this book.

It would be nice if everyone in the world would immediately recognize the value and uniqueness of your idea, and it would be nice if cash offers started pouring in, but as I'm sure you suspect, it's not that easy. The cards are stacked against you in many ways.

Your invention may make great sense to you, but others may not really understand it. You may have overestimated its potential value, and others may believe they can do better. And what about secrecy? How can you even try to sell your invention without someone stealing your idea? What if someone else has come up with the same invention and you just haven't found out about it—can you still get a patent because you thought of it independently?

If you want to succeed with your invention, you have to get the answers to these and many more questions, and you have to get the answers exactly right.

Why We Have a Patent System

The United States has granted nearly 7 million patents, so you know that many others have been through all this before you. Patents and their predecessors go back thousands of years because invention has been a precious natural human resource since long before the dawn of recorded human history.

Strangely enough, most people don't really believe in inventions. These people believe that everything has been thought of before. Some think that a patent is just a legal document that someone went out and paid for in order to get control of ideas that used to belong to everyone.

The reason for these negative views is that there are many intelligent, successful people—including leading scientists, professors, business people, engineers, and attorneys—who just aren't creative. They never really think of anything new, so they don't believe that anyone else has truly new ideas. Granted, this doesn't explain how we got to where we are—walking out of caves and into high-tech homes—but that doesn't seem to trouble people with the anti-invention mindset.

Further, people don't necessarily respect novelty and invention. How many people use an Android phone without being the least bit troubled by the possibility of its features being stolen from Steve Jobs and Apple? What about downloading music without paying for it? It seems like everyone you know does that without worrying about paying the artists who generated the songs they enjoy.

This lack of regard and the correlated tendency to take advantage of opportunities, whether ethical or not, have led to a patent system that fosters progress by offering some protections to inventors.

Defining Patents

Before going further, we need to make sure we all know what a patent is and how it relates to inventions.

An invention is something new that you have thought of or made. The invention may be patentable if it has some use or value in business. The patent is the formal government document that lays out exactly what you have invented and exactly what is protected. There are formal rules, discussed later in the book, about what sorts of inventions can be patented.

A Patent is Not a Secret Document

One popular misconception about patents (I've heard this from professors at UCLA, for instance) is that a patent keeps things secret. In fact, a patent is just the opposite: it is a detailed publication that tells everyone exactly how to copy your invention. Yes, the government requires you to disclose or reveal every detail you know about the invention in exchange for giving you a monopoly to use the idea for twenty years.

Patent Vocab

A patent is a contract between an inventor and the government. The inventor reveals all details of the invention and in exchange, the government provides a right for the inventor to sue anyone who uses the invention without permission.

The opposite of a patent from this point of view is a trade secret. The most famous trade secret is the carefully guarded recipe for Coca Cola. There is no patent on it, and anyone could copy it—if anyone could find out what it was. Trade secrets (as Coca Cola shows) are not limited to twenty years but may remain secret forever. In fact, this is thought to have been one of the original motivations for governments to offer patents—the worry that a citizen would develop a very valuable new process, and then die with the secret so the advance would be lost to society.

Another problem with trade secrets is that it is hard to keep them secret. A few people may need to know the secret and one of them could part ways with you. If that person reveals the secret, you might be able to sue but the secret would be out. All governmental protections that would have been supplied if you had patented your secret would be irretrievably lost.

Someone also could reverse engineer your product and duplicate it without actually learning your secret. After all, you were planning to make and sell your invention, which means that anyone would be able to see what you have done. Patents are likely to protect you from reverse engineering attacks, but trade secret law offers little protection.

A Patent is a Technical Legal Document

It might be nice if you could just write out the details of your manufacturing process, your new machine, or your new chemical compound, turn it in to the government, and get it stamped as a patented invention. That is pretty much how copyrights work. Patents are different.

Copyrights protect originality of expression. If someone produces something, usually writing or art, that is similar to your copyrighted material, you can stop it from being sold. The protection runs for a lifetime and even for fifty to one hundred years after you die. Copyrights originally focused on the printing of the written word but have greatly broadened with the advance of technology to other media.

A patent, however, is far more formal. It draws on aspects of the law of contracts and it draws on a long, complex legal tradition of patent drafting (the formal term for patent writing). Tempting as it may seem, you are unlikely to succeed if you write your own patent unless you are a patent attorney who has passed the patent bar exam.

Patents Expert Pointer

The quick, easy, and inexpensive way to get the patent process started is to file a provisional patent application, which provides one year of instant preliminary protection usable worldwide. This proves your priority and places all the details in the hands of the government with complete confidentiality but is inexpensive and doesn't require a patent agent or attorney.

You can put the process in motion, though. You can write out your idea and file it as a provisional patent application without hiring a professional (see Chapter 7). However, your only chance of getting the patent granted requires that you hire a professional.

The reason for this is that patent law is an incredibly complex thicket of rules and regulations. Worse, patent law is a moving target. From one point of view, much of the current US patent law went into force in 1952 and was updated by the America Invents Act of 2011. However, every legal decision in every patent case (about 2,000 patent litigations every year and thousands of decisions about granting applications) can have an effect on the law. An excellent patent drafter not only knows every nuance of the law, but also has enough experience to anticipate some of the changes that lie in the future.

Patent law is federal law in the United States. This means that individual states in the United States have no input. Only the United States Patent and Trademark Office (USPTO) can grant a patent, and any litigation is held in federal court.

A US patent is good only in the United States. Patent laws are still national laws. You need a separate patent in every country where you may want your patent to be in force. If you have a US patent only, anyone can infringe it freely in Mexico or Canada—as long as no one tries to sell the infringing products in the United States.

Inventor Beware

A patent is only good in the country that grants that patent. There is no international patent and there is not even a European patent; each country has a separate patent-granting process. The inventor has to make cost-benefit choices about which countries to pursue patents in.

Patent law varies from country to country. The European patent system is partly at the level of the European Union and partly at the level of the member countries. You might choose to pay the cost of getting your patent granted in Germany, for instance, but save some money by skipping Luxembourg. Japan, Canada, and Australia have systems similar to the United States or Europe. Once you start to think about China, India, Russia, or the Arab world, however, it gets difficult to even consider extending the reach of your patent. International patents are covered in more detail later in the book.

The Theories Underlying Patents

There are three competing big ideas that explain why we have patents:

1. Natural law: Since you made the invention, you own it. It is your property—your intellectual property. No one should be able to just take something you made. You have an inherent right to your idea.

2. Utilitarianism: Society will benefit if it offers rewards to inventors. Due to rewards, more people will take the time and trouble to invent things. The country with the best rewards will end up with the best technologies to compete against other countries.

3. Nurturing new ideas: Businesses will be able to test new ideas, attract capital investments, develop markets, and perfect products that will advance their business and move technology forward—but only if the government provides a "reverse tax" to subsidize the cost of development. Patents achieve this by providing a monopoly that allows business to charge extra for a new patented product once it reaches the market.

The Role of the Patent Office

The purpose of the USPTO is to grant patents. From this point of view, the USPTO acts like a book publisher: if you meet all the requirements, it organizes the patent form and publishes it for you.

Your patent will come under intense attack, however, if you try to use it to stop an infringer, so the patent office does everything it can to help assure the quality and strength of your patent. To accomplish this, the USPTO assigns a patent examiner, most likely a government specialist who is not a practicing scientist or inventor himself, to review your patent.

The examiner tries to determine if your invention really is new and whether the patent application prepared by your patent agent or patent attorney meets all the requirements. The examiner looks for other inventions in similar fields, unintentionally ambiguous language in the patent text, and omissions in the technical patent claims.

Hopefully the patent examiner's work is thorough and well executed because if you ever file an infringement suit, your opponent could be represented by a patent litigator with twenty years of experience, earning $5 million a year, looking to tear apart every detail of your patent.

At present, the USPTO has a small backlog of over 700,000 patent applications awaiting review. Getting a patent is not fast and it is not easy and it is not necessarily cheap. The process takes two or three years on average and may involve many rounds of corrections and revisions. Your cost (mostly for your agent or attorney) can range from $5,000 to $500,000, depending on the complexity of the patent application, the number of problems that are turned up by the examiners, and who you hire to do the drafting and the prosecution (shepherding the application through the patent office until it is granted).

Overall, the process works well, and most patents that are granted are thorough and meet general standards. Only a tiny fraction of patents ever end up in litigation against infringers. It may be that most patents are so convincing that no one infringes.

Among those that end up in litigation, about half get invalidated by the court even though the patent office granted them. Once a patent is invalidated (unless reversed on appeal), it is as if it never existed. The USPTO gets overruled all the time. For these reasons, the tougher and more aggressive the exam by the USPTO, the better off you are if you make it through to grant.

Benefits of a Patent

Everyone has heard of inventors who make a fortune by developing their invention into a business or even into a new industry, or by winning a fabulously valuable patent infringement lawsuit. This is an anomaly, like becoming the author of a bestselling novel or a rock star—there are a million people who sell only a few books to relatives or who play rock and roll only for their local community center for every one who makes it big.

For many inventors, the best thing about a patent is the certificate they can hang on the wall. It is an impressive statement by the United States government, and it is an important recognition of accomplishment and creativity.

If you work for a company or a university that takes ownership of your patent and doesn't reward inventors—then the prestige and recognition of a certificate is the only benefit you will ever get. If your job is to make inventions, your reward will just be to keep your job.

Consider the most famous and high-earning individual inventor in recent times—Gary Michelson—who won over $1 billion million in a complex patent infringement suit. He spent $62 million in legal costs over fifteen years in a conflict covering six hundred inventions—and he could easily have lost completely at the last moment. The flip side are inventors who spend a couple of million dollars to win low-cost patent infringement suits and get paid damages of just a few thousand dollars.

The most reliably valuable aspect about a patent is the possibility to create a business around it. Yes, you can sell off your patent to a firm that will promise to develop it. Yes, you can license or assign your patent for money or royalties. However, it is difficult to fully establish the value of your invention if you don't personally work to make your invention into a success in the marketplace. For this reason, a patent filing is often part of a business plan. However, if you can convince others of your invention's potential in advance, then licensing, selling, or assigning your invention can be lucrative.

In the end, it is the inventor's job to get the patent and make the invention valuable. No one cares about it the way you do.

The Role of the Courts

Another role of a patent is that it gives you the right to sue. It is true that people generally obey the law because they believe it is right to do so and not necessarily just because they fear punishment. This does work with patents to some extent. However, the commercial world where patents have their impact is about money, and the fear of litigation is a major part of what gives potential infringers pause to think.

Access to the courts is expensive. Inventors can protect their inventions in court if they have built successful companies or have considerable income from their patents. There are also businesses and entities that actually buy patents for the purpose of launching litigation or that invest in patent infringement lawsuits. These firms help level the playing field. A potential infringer who may judge that you do not have the resources to fight a lawsuit, may think twice knowing you can find help.

All in all, the federal laws that define the patent system, the examiners of the USPTO, and the actions of the courts have resulted in a fair and effective system that has helped drive 220 years of world-leading technological innovation in the United States. Nonetheless, to navigate this system effectively and to apply it most beneficially to your invention, you definitely need to study and understand the patent system and its laws.

 **History of Patents**

In This Chapter

➤ Inventions in ancient times

➤ The emergence of the patent system

➤ Famous inventions and inventors

We wouldn't be where we are today but for the inventions of years, decades, centuries, and millennia past. We have a rich history of great new ideas changing our daily lives step by step. That history has led to the formation of the patent system.

Through the Ages

The steady, slow build up of technological advances that has led to our modern world reaches far back into the dawn of prehistory. The Acheulean ax of *Homo habilis* from 3 million years ago probably marks the first widespread relic of a technological advance: the breaking of a stone that could be held comfortably in the hand with the round part resting in the palm and the sharp, jagged edge, or point, forming the working end.

The arrowhead used with a bow and arrow, which is found among *Homo sapiens* populations from South Africa to South America, is a bit of technology that probably predated the spread of our species across the planet 80,000 years ago. It is fair to consider that among warring tribes, those populations that did not have the technology of the bow and arrow were probably overtaken by populations that had them.

The wheel, the domestication of the horse and the cow, the development of a grinding wheel that could turn grain seeds into flour for bread, the discovery that planting rows of seeds in the fall, would lead to fields of growing food in the spring are all critical steps underlying civilization. The control of fire, the cooking of food, learning that metals would liquefy at high heat and could be poured into molds, the advent of thatched roofs, and the cutting of stone to make building walls—each of these basic steps of civilization must have had its start with a flash of genius, a moment of insight and clarity.

By the time that writing emerged in the Middle East, China, and South America, there were thousands of devices and machines in use, as well as large numbers of effective medications.

Ancient Greece and Rome

Probably the earliest famous inventor is Archimedes of Syracuse. The Archimedes' screw was a pump that helped transform irrigation and agriculture in the ancient world. The screw reached the inner edges of a cylinder inside of which it turned. He is also credited with assembling a set of mirrors that focused the sun's heat into a ray that set fire to attacking Roman ships, and a claw/lever system that could flip Roman ships into the air and sink them.

Archimedes did not believe in holding rights to his inventions, but it is clear that he kept his military technologies for his city and certainly did not share them freely in advance with Romans. In fact, the Roman commanders on the ground realized that the way to end the war with Syracuse was to kill Archimedes. This they did and thus the war ended.

It is difficult, even impossible, to identify any well-known Roman inventors because emperors had inventors executed so that existing businesses would not be disrupted by new technology. By the first century AD, the advance of technology in Rome had come to a halt and the last centuries of Rome were marked by technological stagnation.

Inventor Beware

Inventors are not always well liked and appreciated. Many Roman emperors believed in executing inventors to protect the stability of the commercial system. While you think you are introducing a great advance to the world, others may see your work as just an effort to take something away from them.

The Muslim Nations

The Muslim world has similarly been troubled by the position that Sharia law and the larger body of elements out of which laws are derived in Arab and Muslim nations does not support a concept of intellectual property. This turns on the principle that property rights should be applied only to tangibles such as fruit or a building, not to ideas.

Although it is possible to point to medieval innovations in the Muslim world, particularly in mathematics, the output of inventions by the large, well-educated population of a billion Muslims living in Muslim nations is astonishingly small, and patent systems have not played an important role in the economic development of modern Islamic nations.

Europe

With the fall of Rome, the Dark Ages of early European Christianity led to a long period in which thousands of innovations of the ancients were lost altogether. Out of this situation, the progress of technology appears to resume in the fourteenth century when at age nineteen King Edward III of England, in the fifth year of his long reign, did something that was visionary.

At this time, the monarch often granted monopolies to political allies to reward and enrich them. Edward granted a monopoly to John Kempe in 1331 to encourage him and his company to import and develop new weaving technology that had been developed in Flanders (Belgium).

King Edward the III is credited as being one of the greatest military leaders of England: he reversed the course of war and produced a series of victories against France and Scotland. With the grant of an open letter (letter patent) to John Kempe, he also ushered in a critical political innovation.

This critical political innovation was the establishment of using the power of the state to grant monopolies to attract an individual with a new technology. The technology would then enhance the financial strength of England at the expense of its neighbors. Within a decade, similar grants were offered to attract others with secret technologies practiced in other countries with the intent that English artisans would gain access to those technologies.

This practice was an important turning point in the English march toward military and technological supremacy over the following six centuries. This is true not only because of the actual effect of bringing in weavers at the outset, but also because this widely respected and admired king, who reigned for fifty years, had staked out a revolutionary position that modern Americans take for granted: the state should promote the accumulation of technological advances as a means of enhancing the strength of the nation.

Other kings followed Edwards III's lead and continued granting monopolies for technologies throughout the following three centuries. Other European commercial centers followed along as well.

In 1474, Venice established a formal statutory patent system (a system of laws) granting a ten-year monopoly for innovations.

In England, the use of "letters patent" was caught up politically in controversies of the use of royal monopolies in general. James I of England was forced to cancel all monopolies that had been granted, but he prevailed in convincing Parliament that those monopolies granted to inventors for new advances should not be cancelled. It was widely accepted at the time that the favorable treatment of new technology had been extremely beneficial to England over the course of the previous three centuries.

Patent Vocab

A monopoly is a restriction enforced by a government that allows only one entity to do business in that area. In a patent system, monopolies are given to the one true original inventor of something that can be sold or made, or of a method that can be used to make a commercial item.

The resolution of the monopolies problem led to the Statute of Monopolies of 1623 just before King James I's reign drew to a close, and it is the first formal English legal patent system and is the basis from which all subsequent English and US patent laws eventually grew. It took the position that in the future no monopolies could be granted unless they are for completely novel inventions. It created a grant of monopoly lasting fourteen years that was available only to "the true and first inventor."

Elements of the 1623 statute were still in force in England despite major revisions and expansions of patent law in 1843, 1863, 1883, and 1948. Only with the Patents Act of 1977 did the UK formally abandon the 1623 statute completely to harmonize with the patent law of other European countries. Most importantly, this shifted the system to a "first to file" instead of the "first to invent" system that was in force in the United States until it was partially abandoned by legislation in 2011 (see Chapter 6).

The Lessons of Johannes Gutenberg

One of the greatest inventions of all time is the printing press, which led to the wide availability of books. It is hard for us to even imagine a world without books, but until 1455, every book or scroll had to be copied by hand, letter by letter, making books extremely expensive and rare. The story of the invention illustrates a number of aspects of the inventing process as well as pointing out the role of patents.

Making Coins and Souvenirs

Johannes Gutenberg was trained in the craft of making coins. To do this, the gold or silver would be heated to a liquid, and then poured into a disklike mold. The craftsman would engrave the markings of the coin onto a punch. The punch would then be hammered onto the face of the coin, and the soft metal would flow a bit so that the coin surface would have the reverse image of the engraving on the punch.

This method of manufacturing coins was quite ancient even then. Craftsmen learned the method just as their ancestors had before them for thousands of years.

Gutenberg wanted to make a private business based on mass producing something like a coin. He hit on the idea of making souvenirs for an annual religious pilgrimage that was drawing over 100,000 people each year to the town of Aachen in Germany. He made numerous copies of the punch engraving for this souvenir with a plan to line up scores of these punches on a board. He would then use the workings of a wine press to drive all of the punches into a sheet of soft metal at the same time, making dozens of souvenirs in one stroke. Soon he had thousands to sell.

The plan was foiled by a plague, which caused the pilgrimage to be cancelled. Gutenberg was left with the now-useless wine press and metal, which had cost him a great deal.

The Printing Press

Gutenberg's disappointment was followed by a flash of genius: he could engrave a different letter on each punch, then use the press to make them all apply ink to a page of paper, which would lead to the mass production of books. But he had a series of technical challenges to solve.

One challenge he faced was how to extend the letters from each punch after the fine lines were cut. His solution was to engrave molds of each letter and pour the metal so that each letter would be like a punch, but raised rather than indented. He had to experiment with metals to find a mixture that would pour well into the fine lines of the letters, yet hold up when pressed against paper many times. He had to find a paper that could rapidly absorb the ink, and he had to find an ink that would adhere to the letter blocks—the movable type—without running.

The type blocks were movable so that a small number of blocks could be rearranged to make the words on a given page, print numerous copies, and then make up the words of a new page.

The whole development process lasted two decades, consuming money from his family and from various partners, all along keeping the nature of the project secret—even from some of the investors. Eventually, the cost of investment became extremely large, yet he had no product to sell and no proof that the whole thing would ever work.

He finally began to experiment by printing simple schoolbooks referred to as the Donatus, but they had many flaws. Gradually, however, the technical problems were solved and a reliable working process was developed.

When the process was finally honed, Gutenberg printed his first commercial product: a mass-produced Bible. Just as the Bible was ready to be released, his partner Johann Fust called in all the loans and took just about everything Gutenberg had—a judge allowed Gutenberg to keep his home and one older press. Fust and a partner then printed the Bibles, listing themselves as the publishers, not even allowing Gutenberg's name to appear anywhere on the Bible. Others immediately understood what had been done; the secret was out and others immediately began to print books.

Because Gutenberg had no intellectual property rights, because there were no patents, all of the investment, all of the genius and passion, all of the wondrous benefits the world would see in the 450 years between the first Bible and the first e-book were of no monetary value at all for the inventor.

The story highlights some of the important ways that inventions are identified. There was clearly an enormous unmet need for books. The fundamental elements of the technology were there for anyone to use. The printing press could have been built in ancient Greece or by anyone over a 2,000-year spread of time, but it was one person, Gutenberg, who had made it happen. His work demonstrates a key feature of inventions called nonobviousness.

Gutenberg brought together a wine-making machine, a new twist on engraving and coin making, and the ancient tradition of copying texts—three separate areas of technology—to make something that was greater than the sum of its parts.

Simply putting a wine press, a coin punch, and a copied biblical scroll next to each other would not accomplish the synthetic development of a new technology—the printing press of movable type. One can safely say that nearly no one else in the world who saw those three objects next to each other would think to put them together to mass-produce books; it's not an obvious use of those objects.

Patent Vocab

Most patent laws require that an invention be "nonobvious." It must significantly advance what is currently known in a way that would not occur to a knowledgeable but noncreative person.

Thomas Jefferson, Benjamin Franklin, and the US Patent System

The drafters of the US Constitution enshrined the role of a patent system in US law from the outset to "promote the progress of science and useful arts by securing for limited times to authors and inventors the rights to their respective writings and discoveries." Prior to this point, each colony had a rudimentary local patent system, but the new Constitution led to the first US federal patent statute in April of 1790 and at the same time, led to the end of the individual patent systems of the various states.

The rules called for the Secretary of State to take responsibility for the new patent system, and in 1790, the new Secretary of State was Thomas Jefferson—drafter of the Declaration of Independence and future third president of the United States. Jefferson and his colleague Benjamin Franklin were fairly prolific inventors, but both were biased against the idea of patents and monopolies.

The Lightning Rod

Franklin's view is best captured by the success of his most famous invention—the lightning rod. At considerable personal risk, he carried out scientific work that involved using a kite in a storm to prove that lightning was electricity and that an electrical charge could be stored in a battery. This led him to the idea of the lightning rod.

If you placed a metal rod on the side of a building that was tall enough to be the highest point and connected the rod to the ground with a metal wire, you could prevent lightning strikes from burning down buildings. This was a novel and nonobvious idea, and was made available for use to the general public. Most anyone could manage to put up a lightning rod very inexpensively, and no special manufacturing was needed. Franklin had no interest in making or selling these rods.

Implementing a Patent Review Process

Jefferson was also anti-patent, but he took his job seriously and personally read and evaluated every single application submitted in the first three years of the US patent system. He believed there would be numerous frivolous applications that were of little value and others that would merely interfere with industrial progress by casting monopolies across major areas of work through special treatment for various small improvements. He was surprised to encounter a surprising number of really great and original ideas.

By 1793, Jefferson had changed his mind. He then believed that if a rigorous review process could be put in place to weed out worthless and harmful patents at the outset, then the whole process might be very useful.

For a number of years, there was little progress with this idea, and many of the frivolous unevaluated patents he disliked were indeed granted. But in 1836, the US patent office was established and the first patent examination system was put in place.

The World Patent System

Two of the important components of a global patent system are the Patent Cooperation Treaty of 1970 (PCT) and the World Intellectual Property Organization (WIPO), a United Nations entity established in 1967.

From an inventor's point of view, the PCT system is very valuable. It provides for filing a single patent application that can then be used as a source document for a national patent in numerous countries around the world.

After a PCT application is filed, the PCT office carries out a preliminary patent search, which is also available to any country for subsequent evaluation of the patent. However, the process stops after the initial search. This solves the problem of proving the original filing date in every country and allows a filing in one language to be acceptable for establishing the filing date in any contracting state (144 countries are signed on).

The WIPO, on the other hand, is an inventor's nightmare. Like other United Nations bodies, each of the scores of small, poor, or technologically limited countries that are UN members have votes equal to that of the United States—one country, one vote. Most of the members do not produce any inventions or patents and are primarily interested in being sure that they can take and use any invention, software, or publication without expense to their citizens. It is a sort of anti-patent organization. Its history goes a long way toward explaining why there is no such thing as a world patent.

Substantially in reaction to the anti-inventor stance of the WIPO, a separate international patent organization was established by countries with developed patent systems and advanced economies. This grew out of the World Trade Organization (WTO) and is called TRIPs for Agreement on Trade-Related Aspects of Intellectual Property Rights. TRIPs is essentially an undertaking by the member countries to try to harmonize their patent policies and laws.

The Public's View of Great Inventors and Inventions

For most Americans, the prototypical great inventors are Alexander Graham Bell and Thomas Alva Edison. Bell had scores of patents in many fields; Edison was granted more than 1,000 patents. However, it is the telephone and the electric light bulb, each a device that has had significant personal impact on every American, that really draws the public's memory.

The fact that Bell got the patent for the telephone because his filing at the patent office arrived a few hours before the filing by Elisha Grey is not really in the public consciousness. The fact that there were many light bulbs invented before Edison patented his version is also not really in the public consciousness.

Both of these men are perceived as important inventors because of the commercial and cultural impact of what they each created—a technology and a major business bearing their name. Fame and celebrity are clearly part of the perception as well.

Great Simple Inventions

Complexity is not necessarily important for a great invention. The opposite of complexity is demonstrated by an invention by Americus Calahan in 1902: the window envelope. This is just a hole in a piece of paper. The envelope was known, holes were certainly known. However, when you consider how much work the window envelope saves and when you consider how many non-window envelopes were addressed in the hundred years before Callahan's invention, you see that complexity is not at all the same thing as nonobviousness.

Another painfully simple yet far-reaching invention is the Post-it note, which is no more than glue on a piece of paper. This is particularly striking as a simple invention because note pads were already stuck together in a stack by a sheet of glue. Stephen Silver at 3M invented a minimally sticky reusable glue in 1968 but never found any good commercial use for it. Art Fry, also a 3M employee, developed the Post-it note in 1974 after hearing a seminar by Silver.

Patents Expert Pointer

The best proof that an invention is novel and nonobvious is that it is useful and no one has thought of it before. It doesn't matter how simple or complex the invention is as long as it passes this test.

Amazing Complex Inventions

Among the most complex inventions attributed to a single person is the CT (computed axial tomography) scanner invented by Godfrey Hounsefield. He did have the financial resources of a large corporation (Electrical and Musical Industries Ltd., EMI, the Beatles' record company), but his project was so far afield from the company's existing businesses that he was truly working on his own.

The invention is so complex that it is difficult to imagine that only one person was responsible for it. Also remarkable is that it did not seem to fall into a pattern of incremental progress.

Housefield solved one critical problem in the physics of X-ray beam detection, but had spent a great deal of time thinking about how the various parts of the entire system would have to work. Once he had his physical solution, he was able to deploy the resources of EMI to rapidly design and build a series of complex components, and assemble them into a commercially salable scanning machine in a remarkably short period of time.

This was also a personal effort because the medical radiology field did not perceive any need for what Housefield was able to make: cross-sectional images appearing as slices of the brain. In general, people are looking for incremental advances and do not necessarily have the vision to expect or want something that is abruptly different from what they are used to.

Penicillin and the Value of Patents

The story of penicillin and the penicillin patent is often recited to justify the importance of patents. Just as often, the story is misstated by writers who don't understand the patent process and who make the opposite point drawing on the same original facts—asserting that patents are harmful.

The simple versions of the story are that Alexander Fleming discovered penicillin in 1928, but millions then died needlessly because he decided not to patent it but rather make it freely available to the world. With no patent, no one could be bothered to actually produce and sell it.

Others counter that when a patent was issued for a method of mass-producing penicillin in 1948, there had already been millions of doses administered. The anti-patent view lauds Fleming for giving away the invention without trying to become wealthy. The message of the story becomes clearer when considering all the details.

Fleming accidentally discovered penicillin when some mold got onto a bacterial plate in his lab. When he noticed the die-off of the bacteria around the mold, he came to the conclusion that the mold had anti-bacterial properties. However, a number of others had made similar discoveries, including the specific discovery that *Penicillium* was the responsible mold.

Fleming was able to identify the mold species as *Penicillium notatum*, but he then spent twelve years with little or no funding and failed to show that it could be a useful medicine. By 1940, he abandoned all work on penicillin and no medicine was produced. With these facts, it is hard to make a big deal about how he saved the world by not patenting or commercializing.

What did happen next had nothing to do with Fleming himself. The scientists who made penicillin into a drug were not even aware Fleming was still alive. However, Fleming's paper did inspire them.

Howard Florey, Norman Heatley, Ernst Chain, and Edward Abraham at Oxford University solved the chemical structure of the active molecule in penicillin and proved that it was an effective killer of bacteria. In 1941, Florey and Heatley convinced the US government to fund a crash project to learn if the compound could be produced in bulk to use as a medicine, and they went to the United States to work with scientists in Peoria, Illinois.

In 1941, Andrew Moyer, a US scientist who had worked with Heatley, solved the mass production problem. Moyer filed the penicillin patent. Based on the patented method, it was possible to rapidly produce 2.3 million doses in time for the Allied forces to carry this cache with them during the 1944 invasion of Normandy.

It was extremely difficult and expensive for scientists to produce penicillin before Moyer's invention—each dose cost over $20—so it wasn't practical for mass use. The patented method resulted in the mass production of millions of doses, dropping the price to about 50¢ per dose.

The patent issuance was dated 1948, after the millions of doses were produced (the method was kept secret by the military during the war), but they were produced by the method that Moyer had invented in 1941 and for which he eventually received a patent.

So while it is true that Fleming discovered penicillin, didn't patent it, and was ultimately responsible for saving 200 million lives, it is clear that with Fleming's work alone, no one would have been helped. It was the ability to mass-produce the medicine using the patented method that led to the hundreds of millions of cheap, effective doses.

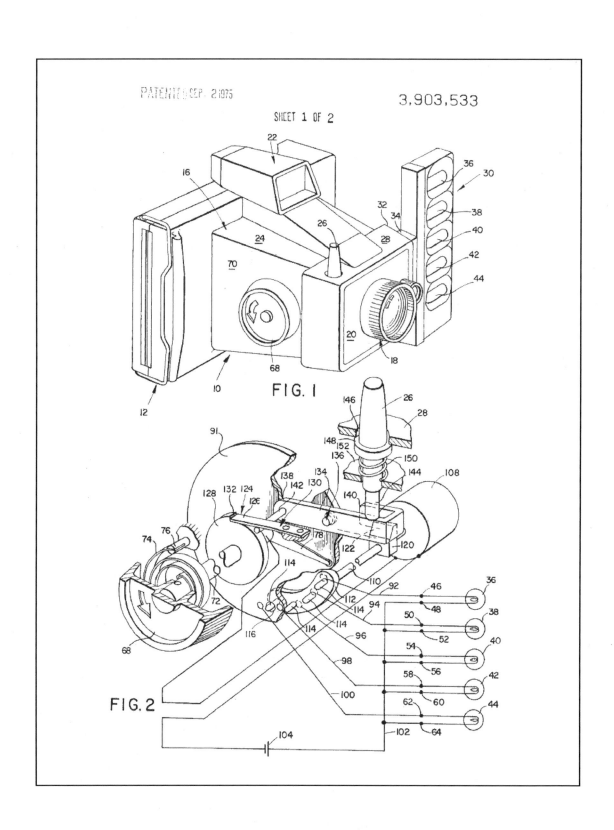

FIG. I

FIG. 2

CHAPTER 3

 # Is Your Invention Patentable?

In This Chapter

➤ What can't be patented

➤ Fine points of software and business method patents

➤ Patenting a design

Some ideas and inventions are patentable and some are not. Whether your idea can be patented certainly should be worked out before you get too far along with your planning and before your expectations get too high.

The first general answer to what can be patented is almost anything, but there are a number of key points in the word *almost*.

Machines versus Ideas

The original purpose for creating a patent system was to encourage industry. The preference has been to offer patents for inventions of new machines, methods of manufacturing, and articles made by manufacturing processes. This fits the common notion of an inventor sitting in the patent office with a model of a complex new mechanical device, waiting to show the examiner how the new machine works and what it can accomplish.

Ideas

The opposite extreme in patentability has to do with ideas. You can't get a patent on an abstract idea or a principle of nature or physics such as a new law of mathematics or the second law of thermodynamics. Your new invention can include a new idea and it can rely on some math and it can make use of the laws of physics, but those aspects can't be the whole sum and substance of the invention.

Let's say you have realized that global energy use can be reduced by 7 percent if everyone in the world would think about conservation three times a day—at 10 am, 12 pm, and 4 pm—thus averting global warming and saving the planet. You would not be able to get a patent on this idea.

Patent Vocab

Patentability refers to the subject matter of an invention. You can get a patent for a machine, device, or industrial process, but you can't get a patent for an abstract idea or mathematical formula. The list of patentable subjects is constantly being refined by the courts.

You could, however, turn your idea into a patentable invention if you were to invent a device that issued a reminder about conservation three times a day—you would get a patent on the device.

You could also get a method patent on the same invention by patenting a method of raising conservation awareness involving three-times-a-day personal reminders (if no one had ever revealed anything like this before you did).

Math and Scientific Discoveries

Just as with ideas, you can't patent a particular basic mathematical algorithm or formula. However, you could patent a machine that is unique in that it acts under the control of a particular formula.

Suppose there is an existing door latch that opens whenever a person pushes two buttons. You invent a latch that opens only when the number of times you push one button times the number of times you push the other button equals twelve (push one button twice and the other six times, or push one button three times and the other button four times). The only difference between the two door latches is that the new one uses multiplication to operate. This would be patentable.

Some kinds of scientific discoveries cannot be patented and other kinds of discoveries can. The law allows for patents on inventions and discoveries within certain limits. The limits are spelled out in detail in patent law or decided as the result of court cases.

If you discovered a new element (like carbon or silicon or gadolinium), you are barred from getting a patent on it. The concept is that you have found something that already exists and there is no creative contribution. If you were given a patent, you would be taking away from everyone else something that was previously available to all.

Inventor Beware

Some scientific discoveries are not patentable because they only reveal what already exists in nature. The fine line between discovering a gene and inventing a new one is an area of unsettled patent law.

If, however, you combine two elements in a new way—for instance, to form a new metal alloy that has more elasticity and strength than either natural metal alone—you may have created a new "composition of matter," and you can certainly get a patent on that.

Or if you discovered a new process or method for getting the two different metals to mix together, you could get a patent on the method. This could be, for instance, grinding each metal into a powder, then pouring the powders into thick oil to mix everything, burning off the oil, and then heating up the powders to melt them and slowly cooling this mixture at a certain rate.

You could get a patent for a new use of a known element. Suppose you find that when a piece of selenium is placed in a roof gutter, leaves don't stick to each other. Instead, leaves wash out of the gutter with each rain, thereby keeping gutters from clogging. You could patent a method of keeping gutters clear by placing selenium in them or coating the metal or any other way of exposing selenium inside the gutter. This would be a patent on a new use of an element.

Software Patents

If you can't patent a formula or mathematical algorithm, can you patent a software program? This turns out to be a very good question, and it is a question on which various authorities don't always completely agree. This is also an area where there has been quite a bit of change over the years.

You may be wondering how it could be that various authorities don't agree if it is all determined by law. First, obviously, laws are different among different countries. In the United States, software is now usually patentable, but in Europe it is usually not patentable. The World Intellectual Property Organization (WIPO), which gives equal votes to small poor countries along with the wealthier countries, is against patents for software. The TRIPs treaty of international rules on intellectual property is in favor of many types of software patents.

Even in the United States there is a lot of complexity to this question. A patent infringement lawsuit is heard by a federal district court judge, who does many different things and may have a personal bias against plaintiffs (inventors) in all patent cases. The district court judge could invalidate the patent based on a ruling he likes to cite from an older case.

When you think the district court judge has got it wrong, you can appeal. Appeals go to the Court of Appeals of the Federal Circuit (CAFC), and if you don't like what it says, it can be appealed further to the Supreme Court. In the United States, however, the Supreme Court does not always have the last word regarding patents.

At one time, the Supreme Court made a decision against software patents, but when the next big appeal came through on this subject, the CAFC overrode the Supreme Court and decided software inventions could be patented.

Judge Rader of the CAFC is coauthor of a book about patents called *Patent Law in a Nutshell.* In discussing the Supreme Court's decision against patenting software, the book says, "the Court's opinion is laced with self-impeaching reasoning." This slightly polite language from one judge to another means the Supreme Court got it wrong, and we're not going to rely on this decision. The end result was that most software programs are now patentable in the United States.

Patents Expert Pointer

The strongest software patents, which are most likely to be safe from a "patentability attack," are those that actually cause something to be made. This is particularly true if the resulting product is something safely inside the zone of traditional patentable subject matter.

The important test of patentability is that the result of the operation of the software program must have a useful effect in business or industry. The fact that the software does not alter the computer itself is irrelevant. The fact that the program can be written out like a poem is irrelevant. The fact that it has algorithms is irrelevant.

Among other points, the CAFC used a technical argument to settle the issue: a program could be installed as firmware in a ROM (Read Only Memory) chip. In that form, the program becomes a solid material object or machine that when operated produces an industrial effect.

Business Method Patents

Another area of confusion and changing opinions concerns the eligibility of business methods to be patented. In Europe, business methods are strictly ineligible for patenting; in Japan, however, business methods can be patented. The position in the United States has varied across time and is still in flux.

In the early 1900s, business method patents fell out of favor in the United States when a patent on a method to keep waiters from embezzling was judged to be unpatentable subject matter. The pendulum started to swing back toward allowing some business method patents during the 1950s and 1960s. In the past ten years, there have been several rounds of disagreement between the USPTO, the CAFC, and the Supreme Court on this matter.

Each business method patent has to be evaluated carefully to determine whether the method pertains to something abstract or whether it applies to a technological or industrial process. One test is to ask if the method acts like a machine or causes a transformation. In either of these situations, it would be patentable.

A business method that simply provided an investment strategy was judged to be an abstract idea and not patentable. Software used to conduct an online auction (as for eBay) was judged to be machinelike and was deemed eligible for a patent.

The most recent change was in the America Invents Act signed into law by President Obama in September 2011. This law allows the validity of a business method patent that deals primarily with financial transactions rather than with something technological in nature to be challenged more easily than other types of patents.

AIA: Change of Rule

Patents for business methods that result in tax avoidance are now deemed unpatentable. All business method patent applications that primarily deal with financial transactions are subject to more stringent validity tests. An ATM machine cannot be treated as a place of business in a business method patent.

Also, any method for "reducing, avoiding, or deferring tax liability" is no longer going to be granted. The technical grounds for this exclusion is a little different from the patentable subject matter issue. Nonetheless, I wouldn't make any big plans for patenting such an invention.

Patents on Genes and Living Things

In early patent law, only nonliving things could be patented. However, it has been possible to patent certain types of plants in the United States since 1930. These were typically new strains of plants with particular new desirable properties or crossbreeds such as the nectarine.

Gradually, with the march of technology the patentability of living things has broadened. At present, recent court decisions have permitted the patenting of a type of mouse in which genetic engineering was used to include certain useful new features. However, there is an exclusion for the patenting of humans. In fact, if any substantial part of the genome of a new type of animal is human, there may be a ban on patenting it.

The law on gene patenting is in active flux. A major fundamental case on the patenting of genes by Myriad Genetics is in the midst of review and appeal as this book goes to press. The district court judge has ruled that a patent based simply on purifying and identifying a gene (in this case, a particular gene that marks the potential cancer risk a person has) is invalid on the grounds of subject matter. The CAFC liked some of the arguments but not others.

It seems clear that modifying a gene to accomplish a new desired effect will probably remain patentable. The situation on decoding and identifying an existing gene is still uncertain.

Medical Use Patents

It is true that the courts do not want to be put in the position of enforcing a patent in such a way that doctors are prevented from taking care of their patients. However, potential inventors and investors would be driven away from medical technology if patents for new technology in medicine and surgery were not allowed.

So the United States does allow patenting of medical and surgical technologies and methods, but it limits the damages that can be sought from doctors who are infringing on a patent in order to treat their patients. This balance encourages medical innovation and patents but helps keep some distance between the patent courts and the operating room or clinic.

In Europe, the rules are more restrictive: you can't get a patent for a method of doing medical or surgical work. However, European law does allow patents for products (such as pharmaceutical or a particular surgical instrument or implant) made for medical use.

In addition, European patent law allows for patents on the first medical use of an existing object or substance. This appears to allow a method or process patent as opposed to a patent on the substance or product itself.

An example of a patent that falls under this law is a chemical compound previously used for chemical purification that turned out to be a pain reliever, which is what happened in the discovery of the drug that is the basis of Tylenol (paracetamol). You can get a patent for using the compound as a medicine, but not for giving this medicine to patients in a powder twice a day to counteract a fever.

Inventions Involving the Written Word

The difference between a patent and a copyright is explained by a rule that states that something cannot be new or novel from the point of view of patents just because the words on the invention are different from the words on a similar previously patented invention. But new and different words are the very essence of the uniqueness of expression that the copyright law is based on.

There are some situations where the use of words can lead to a patent. An example might be a new way of applying words to an object, for example having the stitching on a baseball spell out the names of the various teams. Another example is of a measuring cup with the measurements in thirds and ninths along with the usual halves and quarters. This is allowable because it changes the use of the measuring cup.

Written music, songs, and speeches also cannot be patented, although a method of using or transmitting them certainly can be. It all works around the idea of industrial use.

Utility Patents vs Design Patents

One further twist is that there is a special distinct type of patent for just the design or appearance of something. A famous design patent is that of the appearance of the iPod. Of course there are also utility patents that cover various aspects of the construction and operation of the iPod, but the design patent covers just the appearance.

Another good example is patenting an unusual-looking telephone from Bang & Olufsen. The design patent prevents someone else from designing another telephone that looks similar to the Bang & Olufsen phone.

Often a design patent can be based entirely on a drawing rather than on a detailed written description as required for a standard utility patent.

There is a difference between something subject to a design patent and, for instance, a sculpture. A sculpture is considered to be a work of art made once or made in a limited number of copies and is subject to copyright as an artistic expression.

A design patent is for an object that is manufactured (an "article of manufacture") for use in commerce even if there is only one made. An example of this would be an unusual display stand sold to the US Park Service for supporting the Liberty Bell in Philadelphia.

For a design to be patentable, there has to be some ornamental quality to it that is not a direct result of the use or operation of the device. It also has to be distinct from previous designs in order to be considered new or novel.

PART TWO

Value and Ownership of Patents

M. S. ROSENFELD.

INDICATING TAIL LAMP.

APPLICATION FILED AUG. 26, 1910.

1,010,806.

Patented Dec. 5, 1911.

3 SHEETS—SHEET 2.

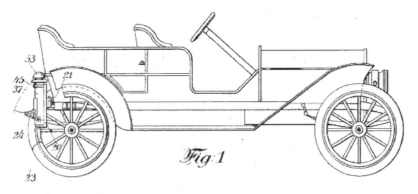

Fig. 1

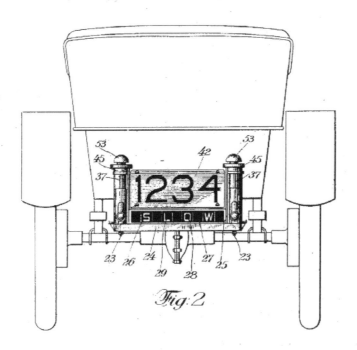

Fig. 2

Witnesses
John E Prayer
Jesse H. Swiedler.

Maurice S. Rosenfeld Inventor
By Attorney

CHAPTER 4

How to Predict the Value of Your Invention

In This Chapter

> ➤ Inventing at the right time

> ➤ Science vs Invention

> ➤ Calculating a patent's monetary promise

An important step in deciding whether to pursue a patent is to decide whether you have a great invention. Recognizing the downsides of your own invention is difficult, and the task is even more challenging for scientists and physicians who are used to participating in innovation and progress but don't always know how to distinguish what makes something a great piece of science as opposed to what makes it a great new invention. Read on for ways to determine the potential worth of an invention.

Will an Invention Catch On Like Wildfire?

One basic determination of an invention's worth is its financial potential. If your invention turns out to be truly new and patentable, and if it is destined to mature into a granted US patent that will withstand any validity challenge by any infringer, how much money can it earn in the world's marketplace? The answer to this question is not always as simple as it seems.

Let's say you have developed a great new method for rapidly charging replaceable batteries in electric automobiles. You estimate that each machine will sell for $100,000 and that there will need to be 10,000 such machines around the country. That might mean that your invention will be worth $1 billion in sales. The catch is that there aren't any electric cars using replaceable batteries yet. What if your patent is granted, but the market doesn't mature until your patent expires twenty years from now? In that case, your invention would be completely worthless and would likely bankrupt you.

What if you have information that convinces you there will be a huge market for your invention ten years from now? How sure are you that no one else will invent something better and different enough to get around your patent between now and then?

You learn that there are ten start-up businesses in your field that already have millions of dollars in venture capital and are racing to develop the best technology. Once again, your patent could prove to be worthless in this scenario. But your luck could change if one of those companies wanted to buy or license your invention to help protect its own investment.

Let's consider a far less dramatic scenario. You are a plumber and you know that one million pipe joiners are bought every year for $10 each. You have invented a much better joiner with a unique and useful feature that you're sure every plumber will love on sight, and it is inexpensive to make. You will be able to sell your invention starting tomorrow, even before the patent is granted.

You know you can't beat the big distributors, but you're sure the invention is so good that it will rapidly let you elbow your way into at least 5 percent of the market. Well that could be worth $500,000 in annual sales, of which 50 percent is profit. The income from early sales will pay for the patent filing costs. There is reason to believe you can make $2.5 million in profits over ten years, and you already have some ideas about additional patentable products you could launch with the profits from your current invention. This may not be a dramatic advance that alters the human condition, but it might be a great patentable product and a great start for a new business.

Two Story Lines: Great Science vs Great Invention

One helpful exercise for understanding what makes a great invention is to compare great science to great invention. These two areas of progress have very different story lines.

In a scientific journal article, the story runs something like this:

Everyone in the field agrees on what has been done so far. Everyone agrees on the next big question that needs to be solved to move the field forward. Everyone agrees on the kind of experiment that's needed, how it should be organized, what equipment should be used, and what controls should be put in place.

We competently perform that experiment and our results are exactly what everyone predicted they would be. The problem has been solved, removing all doubt, and the field can now move on to the next issue.

An invention's story is different:

No one had any idea that something new could be done in the area in question. I have done something that no one even ever thought about doing. I used materials and processes from completely different fields that no one in this field ever even imagined. I have now invented something that is totally surprising that no one ever predicted. It is fantastically useful and solves numerous well-known problems that were wrongly thought to be insoluble. This is such a huge and unexpected advance that no further work will be needed in this new area.

The reason for the difference in these two examples is that science relies on building knowledge steadily within a well-known theoretical framework. An invention, on the other hand, has to be disconnected from what was expected because it must be convincingly novel and nonobvious.

In science, the effective value of an advance or discovery comes from the strength and reliability of the result and from the degree to which it ties together and explains what was known before. The value of an invention is a business issue.

Great inventions meet the requirements of patent law very nicely. This results in successful navigation of the patent examination process through the granting of the patent. It also predicts success if the patent ever has to be litigated. The commercial value of an invention is entirely different.

Inventor Beware

It is important for inventors to become knowledgeable about the business environment and potential market for their inventions. No one wants to spend the time and money getting a patent on something that is a great and novel idea but has little or no commercial value.

Be careful in this judgment, though. Paul Lauterbur decided not to file a patent when he invented the MRI (magnet resonance imaging) machine because advisors told him there was no potential commercial value.

A good example of this would be a device that is built into a hat that can determine the exact shade of blue in the sky at any given moment. This might be completely unique and meet all requirements of a granted patent in terms of patentable subject matter, novelty, and nonobviousness. However, it might be something that no one would ever want to buy.

Further, if five people in the world decided to buy one, they might be willing to pay only 10¢ for it.

A great invention must be both the basis of a great patent, and also capable of having great commercial success. In science, you get a grant, you do your research, and then you publish. With inventions, you could spend $5,000 doing initial development, $100,000 getting the patent, and $3 million in court to stop an infringer. So if you ever contemplate the remote possibility of stopping an infringer, you should carefully consider whether you have a multimillion-dollar invention. If it is worth only $1 million, it might not make financial sense to pursue an infringer no matter how clever and unique your invention seems.

How Scientists View Inventors

Scientists tend to be suspicious and disdainful of inventors. There are several reasons for this. One reason is personal.

The Personal Distrust

A prominent scientist can be considered a great productive genius with a hundred important scientific publications but also be relatively uncreative. The scientist has reliably and repeatedly advanced the field along predictable lines.

It is difficult for such a person to accept that there might be something lacking in his capabilities. Traditional scientists, therefore, suspect there must be something wrong with inventions if they have never invented anything.

Scientists may say that the very idea of invention is false—that all knowledge develops steadily and progressively from one firm point to the next along lines of theory generated by the previous results. The idea that sudden unexpected discoveries even exist is just commercial and political propaganda. People should devote themselves to generating new knowledge for all. Invention diverts funds into businesses that profit an individual when all that money should go for pure science to advance general knowledge.

Inventions are Disruptive

Another reason scientists may dislike inventions is that they tend to be disruptive. This is really a matter of fear. One's carefully built-up recognition and prominence can be turned upside down in an instant if an inventor discovers a new scientific instrument that reveals that ten years of data collected by scientists were actually incorrect and misleading. In science, there are scientific revolutions, but they are rare. Invention seems like a much more immediate and constant threat.

Patents Expert Pointer

Once your patent application is filed, there is no requirement to keep it secret. For an academic, this means you can publish immediately following your filing (as long as you don't disclose anything additional). This is actually a great way to start building interest in your new invention while also protecting your day job as an academic scientist.

The Bane of Secrecy

Finally, there is the problem that patents seem to be related to secrecy, something that scientists like to claim they despise. This issue arises because scientists rush to publish their findings as soon as they are confirmed. Inventions may remain shrouded in secrecy for months or years.

This turns out to be mostly a misimpression. There is no requirement that inventions remain secret as long as the inventor follows the necessary procedures required to safely make the discovery public (see Chapter 6).

It is true that an inventor can keep an invention secret for a considerable time and still succeed with it. However, scientists employ secrecy when they compete for grants and try to get ahead of other labs.

For a corporate or university inventor, these issues can weigh heavily on a decision to report an invention. You may not be the person taking the financial risk, but you may also not be the person getting any benefits. If you are working in the research lab of a large corporation, the more inventions you make, the more of a hero you will be. At a university, making inventions and filing patents could actually lower your colleagues' esteem for your work.

For this reason, academic scientists often avoid listing any patents on their résumé. When an academic comes up for promotion (from assistant professor to associate professor or from associate to full professor) the committee reviewing the scientist's work often places no value on any patents.

Technology Transfer

Universities do often recognize that they have a mission to improve the economy of the state or city or to help in the development of businesses. In the United States, most universities have technology transfer offices that get patents on faculty inventions, and then transfer the technology to interested businesses. This places some value on an academic who produces

Patent Vocab

Technology transfer is a process used by universities to try to move inventions made in academia into the world of business. This is done both to serve the community by making new ideas available for commercial use, and also to try to earn money for the university.

patents, but many universities have found technology transfer to be an unproductive expense and become frustrated with a faculty member who submits too many expensive inventions.

One reason technology transfer often doesn't work out is because companies with their own research departments can have a bias against inventions coming from sources outside their own labs. This is called the "not invented here" syndrome. It is based on the idea among corporate scientists that they understand their own field and its commercial aspects, and that an academic cannot make an invention that is as useful as one they can produce.

There is also the issue that it may be a threat to a corporate scientist's job if the company is buying technology from the universities. After all, why bother to employ scientists and maintain a lab if the important advances are coming from elsewhere.

When it comes time for a company to decide about licensing an invention from a university, it will have no choice but to turn to its own scientists to review the invention in order to advise the company on its value.

Who Owns Your Invention?

If you think it is obvious that you own your own inventions, think again. Invention ownership, and therefore the right to the patent, is less cut-and-dried than you might think. This is especially true if the invention was created by an employee of a private or public institution.

Company or University Ownership

Most inventors do not own their own patents. Anyone who has a job may find that their employer is the owner of their inventions. If you invent something while you're self-employed or unemployed, however, there is no question about who owns the invention—you do.

Sometimes, employer ownership of inventions is easy to understand; at other times it is surprising. One example that makes sense to most people is in the case of a research company that hires a person with a job description and performance evaluations all directed toward making inventions. The company advertised for an inventor and hired the person to be an inventor, so it is no surprise that the employer would own the inventions.

The University as Owner

Let's take this a step further to the generalized idea of a university faculty member whose job is to teach and do research. The professor uses the university's laboratory and does research necessary to stay employed and get promoted.

One day, the professor observes something surprising in the lab and realizes that he has accidentally discovered a new chemical that will be useful and that many companies and researchers will want to pay for. The university will say it owns the invention because the discovery was made during working hours, using university equipment, in the course of the professor's usual work.

In part, this all goes back to some of the theory of patents and inventions discussed in Chapter 1. There is an idea that if a person goes into the forest, picks up a fallen tree limb, and carves it into a beautiful sculpture, then obviously the carver is the owner of the sculpture. The carver can keep it or sell it and keep every penny earned from the sale. This is called the natural law of ownership.

The Case of the Factory Worker

However, then consider a person who works in a factory operating a press to cut parts out of metal sheets, then shaping them and assembling them into metal boxes for sale by the company. Does the worker have any ownership rights to these metal boxes? The answer is no.

The underlying reason an employee in this scenario has no ownership has to do with the job contract in which the employee agreed to be paid in exchange for labor. Not such a bad deal since the employer has to pay the employee even if no boxes are sold or if the metal supply runs out. The employee gets time off, vacation, health benefits, workers' comp insurance, but no ownership of the product. This is part of what Karl Marx called the alienation of labor. The employee makes the box but does not design it, market it, nor sell it. The employee created the box out of raw materials but has no ownership right to it at all .

Employers treat employees' inventions in the same way—they own what employees invent. However, there is a little bit of uncertainty on this issue.

Inventing After Hours

Let's suppose a company hires an inventor to develop new types of electronic capacitors. At home on the weekend, however, the employee develops a new woodworking tool similar to a chisel, but with a new and useful advanced feature. Does the electronics company employer own the chisel invention?

Inventor Beware

If you are employed or if you have received a grant for work that leads to your invention, you may not be the owner of your invention. It may be tempting to invent at home and simply not tell your employer, but that's a very bad idea. In one famous case, an inventor did that and became fabulously successful. However, he later lost everything when he filed a patent infringement suit. The infringer got the case dismissed because only the true owner can sue for infringement. Then, the employer sued the inventor to get its money back when it learned what had happened. It is better to settle the issue definitively at the outset.

Well, you might think that the answer is a simple no; the employee clearly owns the chisel invention. After all, the employee wasn't at work and didn't make this invention during working hours. And no raw materials or laboratory space owned by the company was used. Not only that, but the invention has nothing to do with electronic capacitors, and it seems obvious that the company would not spend a penny on filing a patent on chisels and is unlikely to make or sell any chisels.

However, this is not as clear as it seems to be. What if the employee goes out drinking on Saturday night with friends, and after a few rounds of bourbon has a flash of genius about how to make a much better capacitor? The employee sleeps off the binge, wakes up late Sunday afternoon, is so excited about the idea because it still makes sense, and buys some parts from a local electronics store. The employee builds the improved capacitor on the kitchen table, turns on the power, and it works. It is a huge breakthrough. Does the employee or the company own this new, improved capacitor?

This question is one that could be fought out in court because the fact that the employee was off duty, using personal materials and supplies, does not settle the matter. The employer will argue that all the information and experimentation that brought the employee to the point of this inspiration was bought and paid for by the employer. Further, after paying the employee a very nice salary for a year during which nothing was invented, the employer is not going to just walk away and get nothing in return for its pay and patience.

Assignment Agreements

To avoid the disputes, most employers of scientists and engineers require them to sign an agreement when they start work that puts all this to rest. The agreement seals the deal by assigning any and all inventions they may make to the company. Some assignment agreements are more comprehensive than others, and there is some variation on this issue from country to country.

Patent Vocab

A patent assignment agreement is a written agreement wherein the owner (or potential owner) of an invention/patent permanently gives away all right, title, and interest in the intellectual property in exchange for something of some value such as a royalty or one-time payment.

You may not remember signing anything like this because the day you started work five years ago you were so excited to get your job, that you signed a stack of fine-print employment forms without even reading them. Let's suppose, the human resources person actually did point out that one of the papers was an "assignment of all inventions you may make during the course of your employment." You didn't know at the time that you would ever invent anything, and you were not interested in risking your job offer or angering your new employer by objecting, so you signed.

It is true that signing an important document like this, which you didn't really understand and couldn't reasonably get an attorney's opinion about, could be challenged in court as "unconscionable or unfair," but most courts would probably not support you on this. The reason is partly that it is so standard and that it is just being done to further confirm and document the fact that the employer probably owns the invention even without an assignment.

Finally, there are laws in some countries and prior court cases in others that further set the framework and give the employer the upper hand. Although it is true that the particular details of an individual case may change so that the employee gets awarded ownership, most employees don't want to get into this fight.

Inventing Post-employment

If you think of an invention and don't tell anyone, can you quit your job and then start to document the invention once you are self-employed or unemployed? Yes, you can do that and it will probably result in you owning your invention. Of course, quitting a good job to begin a long, expensive, and uncertain course of applying for a patent might not make a lot of sense when you add up all the pluses and minuses.

This scenario often makes sense for employees who see the impending end of employment such as a scientist who works for the United States government and is near retirement. They won't be at the lab anymore to promote and develop the invention. If they wait until they're retired, they will own the invention and may be able to get investors to help build a start-up company.

The reason this is even possible is that an invention is intellectual property, and at the outset it is just an idea in your mind. But an employer may be able to argue in court that the ex-employee originally had the idea while still employed. This could lead to big trouble for the employee.

The best way to get control of your own inventions is to seek an agreement from your employer that gives you control. Sometimes this is easy, sometimes it isn't.

I once obtained a release when I had a part-time job with the United Kingdom's National Health Service (NHS). It had a harsh policy on inventions (leftover from the pre-Margaret Thatcher era), which included the following:

➤ Employees would be in trouble for inventing anything because they were supposed to be working full-time taking care of patients.

➤ The NHS owned anything invented by employees—night or day, on duty or off duty.

➤ The NHS would not spend a penny to develop or attempt to file for a patent for any inventions—employees would have to provide the money for the NHS to do any patent filing.

➤ If the invention ever generated any income or royalties, the NHS would keep 100 percent of it. Employees already had a good salary and were not entitled to any more money than they were already earning.

Yet, when I asked in writing, the NHS readily agreed to formally sign away any rights it might have in an invention I had made. You won't find this situation with most US employers.

When You Don't Own Your Invention

Once you get past the upset of realizing that your company owns your invention, you may be open to understanding that this might not be such a bad thing. Remember, there can be a great deal of risk and expense involved in filing for a patent and trying to get it granted. Beyond that, even if you succeed with the patent office and get your patent, how will you develop it, and what would you do if a big company started infringing on your patent?

When a Company Owns Your Invention

If your invention works out and makes your company a lot of money, it will probably appreciate your contribution and you will benefit. If there is a cutback in hard times and your company must lay off employees, your boss's belief in the possibility that you have another great invention in you will help you keep your job.

Because of the indefiniteness over who will own an employee's patent, most employers protect themselves by getting an assignment in advance and an additional assignment on the specific invention when a patent is about to be applied for.

In order for the validity of the assignment to be guaranteed in court, the company will probably agree to give you something in exchange for what you are giving it. Often, what you get is listed in an employee manual or other company or university policy document. This may include a variety of benefits, most commonly a percentage share in any earnings after the company has been paid back for the money it spent getting the patent filed and granted.

Different Approaches Universities Take to Inventions

Some universities, such as Stanford, strongly encourage faculty inventors. The result is the explosive development of Silicon Valley, with thousands of startup companies, some now worth billions.

When faculty members are allowed to work for companies while staying on the faculty, the university gets the benefit of having a brilliant creative professor teaching and inspiring students and often gets shares in the start-up companies. The United States benefits by being a world powerhouse of technology generation.

California attracts businesses that want to be in the middle of all of this in order to exchange ideas, employees, and deals. Stanford attracts the best students and new faculty while it makes money from its shares.

Contrast this with UCLA in Los Angeles, which has a patent policy more like the UK National Health Service. There are virtually no startup companies anywhere near UCLA. A professor can be fired for participating in one or even for receiving income from an invention that the university doesn't own.

The policy at UCLA reflects completely different priorities and concerns from those of Stanford. It may have high-paid faculty members, but when a faculty member tries to do outside consulting for additional income, the university sees that work as undermining the morale of other academics, taking advantage of the state salary to do work that doesn't benefit the state, and generally abusing the terms of employment. Inventions are viewed as just another way to evade the salary and work rules. This is because the state-owned university is not a business and really can't necessarily envision how it will benefit from inventions.

Policies such as those of UCLA are why an academic who invents something may prefer to just give the idea to the public without filing a patent, then handing it over to the university.

This may mean that no one ever benefits financially from the invention: without a patent, no one will invest, the product never gets made, and the result is just a waste. However, this approach avoids any risk that the professor will be fired because of the invention.

This approach seems to be bad industrial policy for the United States, California, and the university, but there are often competing policy concerns and priorities that lead to this sort of situation.

Forming Your Own Company

Patents Expert Pointer

The smart inventor turns employer ownership of the invention into a huge advantage. The employer faces the risk and expense of getting the patent written and through the patent office. If the patent gets granted, you will be in a much better position to round up investors, set up a company, and then come back to your employer with an offer to codevelop or take over your invention at this later point in the game.

Even if your employer has an uninspiring policy for rewarding inventors, you can still benefit very nicely if you have the drive to quit your job and start your own business. This is particularly true for university and other government employed inventors.

The basic idea is that the university or government agency has kept ownership and filed the patent application because it knows the invention is valuable or because some higher government policy forces it to try to get patents on inventions made by employees.

However, because the university or agency is not a business, it has no capability for actually generating income from the invention once the patent is granted. It will offer licenses to existing businesses, but most companies prefer to develop inventions they already own.

Your fantastic invention has now been granted a patent and is now one of 237 university-owned inventions listed on the back of a brochure or on a web page as being available for licensing. That's just great, but nothing is happening and no one benefits.

You, the inventor, may be the one person who has the drive, knowledge, and motivation to change all that. You quit your job, start a new company, get some potential investors lined up, and contact the university to let it know you are ready to license your own invention.

This may sound risky, but that is how it often has to be done. You can see immediately that it is much nicer if your employer allows you to keep your job while you set up the startup company, get the license, and win the investments.

Some corporate employers such as 3M have had policies that allow entrepreneurial employees to do exactly this. They let employees set up a wholly owned or partially owned subsidiary that benefits from their energy, inspiration, and drive. This allows a big slow-acting company to compete with nimble startups, by having its own "startups" emerge internally. If you want to stay employed at your corporation, you might want to pitch this concept to the company executives.

Another tried-and-true technique for the employed inventor is to pitch the idea and the invention to a person who has created startups in the past, such as a person who has been a CEO of a tech company but is out of work due to a sale of his company (or due to a bankruptcy/failure of such a company). The ex-CEO can set up the company, line up the potential investors, become CEO of the new company, get the patent license, and then hire you away from your existing job after everything is already in motion.

If you take all the risk by quitting your job and starting the company yourself, you will start off owning the whole company and gradually give away pieces of ownership as you bring in partners and investors. If someone else starts the new company, you may only become an employee or your degree of share ownership and control could be limited. This is the risk/reward ratio that every inventor and entrepreneur has to navigate.

PART THREE

The USPTO: Getting Filed and Getting Granted

PTO

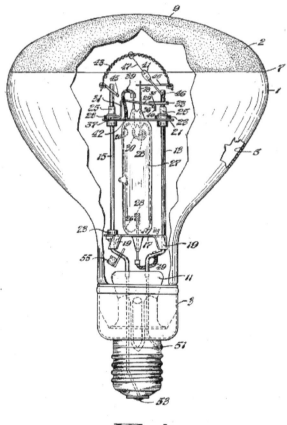

FIG. 1

INVENTOR.

Michel E. Macksoud

BY

Kenway, Jenney, Witter & Hildreth

attys.

CHAPTER 6

 Keeping Your Idea Confidential

In This Chapter

➤ The importance of secrecy

➤ Impact of the new America Invents Act

➤ What to do if someone steals your idea

For many inventors, particularly first-time inventors, the question of secrecy is foremost on their minds. How do you keep someone from stealing your idea? What happens if you discussed your idea with a colleague just before you realized you could patent it? What can be done after a leak to save the day?

The America Invents Act

The rules on secrecy have been very different in Europe than they have been in the United States. Most of that changed, however, with the September 2011 America Invents Act (AIA), which brought the United States more in line with Europe.

The secrecy issue affects several different questions. Firstly, if the colleague you spoke to decides to run to the patent office and file for a patent on your idea—will you get a chance to prove in court that you really thought of it first?

What if you gave a talk at a scientific meeting in front of 1,000 scientific colleagues where you announced to everyone for the first time what you have done? If any of them try to race to the patent office with your idea, can you provide testimony from the scores of witnesses who can swear that they heard it from you a few hours before the other person filed his patent application?

Inventor Beware

The American Invents Act (AIA) of 2011 signed into law by President Obama introduced huge changes in American patent law. It is unwise to make any important choices about protecting your ideas or filing your patent if you rely on guidance from books published before 2012. Even books or web pages written about the AIA before the final law was signed in September of 2011 can be misleading due to last-minute changes in the law.

Interference and Derivation

One result of the AIA is a new type of legal process called a derivation proceeding—the purpose of this is to address the "stolen idea" problem. A derivation proceeding is now available to determine if the person filing for a patent actually got the idea from someone else and is filing without that person's permission.

If it is proven in the proceeding that the person who claimed to be the inventor merely copied what someone else invented, then the derived claims of the patent would no longer belong to the false inventor. Further, it is possible to actually sue someone for stealing an idea in this way if that action damaged the true inventor in any way.

The AIA did away with an older type of proceeding you may read of in other older sources called the interference proceeding. There is a fundamental difference between the old interference process and the new derivation process.

Patent Vocab

In the United States, if you can prove that a person who filed for a patent had copied the idea from you, you have proven derivation. The patent application stays valid, but the patent office adds your name to the application as the inventor and deletes the name of the person who stole your idea.

Under the old interference process, the patent office would be interested in your work if you had conceived of the idea years earlier and worked diligently day by day—keeping it completely secret at all times—up to the point where you were able to file a patent application. If someone else filed an application for a similar invention and could not prove he thought of the idea before you did, then yours would be deemed prior art, which would render the relevant claims of the other patent invalid. You were first to conceive of the invention, so you are the true inventor and you would get the patent.

There was no suggestion of any relationship between you and the other inventor. You both happened to have the same idea, but you could prove that you thought of it first. You would present evidence to the patent office and the patent office would conduct the investigation.

In this old system, you won if you could prove you thought of it first even if the other inventor thought of it independently. You won whether or not the other inventor filed before you did.

In contrast, the derivation process only seeks to learn if the person who claims to be the inventor actually learned of the invention from you and filed without your permission. If that is proven, then the name of the inventor on the patent gets changed, so you are listed as the inventor. If your competitor thought of it independently and gets to the patent office first, however—you lose.

You can start the process by providing evidence to the patent office, or you can file a claim in federal district court whether or not the patent office agrees that there is reason to investigate. The derivation process can be more expensive and involves a wider array of evidence gathering.

In the interference process, a patent owner could initiate an interference proceeding in federal court many years after the patent was granted. Timing is now much more critical. Derivation proceedings have tighter rules that require an inventor to speak up within one year of the patent office publishing the offending claim if he or she wants the patent office to conduct the derivation proceeding, or within one year of the date the offending patent was granted if the inventor wants to litigate it in court.

First to File

Under the new AIA law, it doesn't matter how long you've worked in secret or how many people you've told about your invention; all that matters is you file the patent for your idea before anyone else does. Here's a scenario:

You worked in secret in your basement for three years and have been trying to raise money to hire an attorney to write the patent application. On the other side of the country in a huge corporate laboratory, a scientist has the same idea you first thought of three years ago. He calls the in-house patent counsel who rockets in the patent filing the next day. The corporation gets the patent. You lose. End of story.

It no longer matters in the United States who has an idea first as long as it wasn't copied

AIA: Change of Rule

Forget about all the advice on writing down your ideas and getting the notes signed by a witness. The law in the United States is no longer concerned with the question of who thought of an invention first. All that really matters now is which inventor files first at the patent office. Giving a scientific presentation to get priority and then having a year to decide whether or not to file is a thing of the past. The only way to get priority is to file your idea at the patent office before anyone else does.

from someone else.

Under old US patent law (pre-2011), you could reveal your idea at a meeting, and then spend a one-year grace period deciding whether or not to file a patent. You can see the problem with this under the new AIA rules.

Suppose someone hears an audio recording of the meeting, then goes to work and tells a scientist about your great idea as if it were their own. The scientist works out the details, calls the in-house attorney, and files a patent. How will you ever prove that this interloper learned of the invention from you? You have your suspicions, but proof may be elusive. The company filed first, and you can't prove derivation.

Under the old law, the patent office could ask the company to produce laboratory notebooks proving that it had conceived of the invention before you gave your talk. Under the new law, you have to prove the company got the idea from you directly. The fact that it conceived of the idea later is irrelevant as long as the company filed first and you can't prove derivation.

In theory the grace period still exists, but many attorneys believe that it will be too difficult to safely rely on it under the new law.

The History Behind AIA

How did this change in patent law happen? Why would President Obama sign in such a law?

This change was made under a belief, which many feel was mistaken, that we must harmonize with a better, newer European system. In fact, as explained in Chapter 2, our laws were derived from the British 1623 Statute of Monopolies. The Parliament forced King James I to accept that a patent would go only to the true first inventor so that there would be no way for the king to get around the law and give patents to wealthy associates.

Other parts of Europe had their own issues with patent law. There was no unified German patent law until 1877. When it was set up, it was a first-to-file system so that the big corporations in Germany had the advantage from the start. Overall, this has helped empower the large German corporations, but it has never been successful at promoting individual inventors and startup companies. The French had a chaotic system that no one was satisfied with.

European vs US Patent Law

When the European patent law was laid out, it followed the German model. The British agreed to change to the German-style system in 1977, but the United States had resisted—until 2011. The old US model was called First to Invent, or FTI. The European model is called First to File, or FTF. The new US version is called First Inventor to File, or FITF.

The Europeans had three major criticism of the older US system:

1. They believed it encouraged conflict and litigation. In a European-style First to File system, if one patent is filed ten minutes ahead of the other, all the past history is irrelevant. The first one to arrive wins. It's very simple.

2. The Europeans were at a disadvantage when they tried to file in the United States because the United States refused to consider evidence of earlier conception unless the inventor was physically in the USA when he had the earlier conception. Even a US inventor working in England, for instance, was not allowed to provide evidence of earlier conception in the case of an interference proceeding in the United States.

3. The Europeans had wider support for their system and felt that the inconsistency stood in the way of a future uniform worldwide system.

You might wonder if the purpose was to be consistent, then why not go completely over to a European FTF system; why opt for something in between, the FITF system? The answer is that what the Parliament forced James I to accept in 1623 in England ended up enshrined in the United States Constitution.

Article I, Section 8, of the United States Constitution says: "To promote the Progress of Science and useful Arts, by securing for limited Times to Authors and Inventors the exclusive Right to their respective Writings and Discoveries." The Congress in 2011 thought that the patent law change could be declared unconstitutional if it did not show an effort to make sure that the patent was going to the actual inventor.

If you assume that the derivation process will work and will stop any copying of ideas, then in the situation of two independent inventors, both are inventors.

To solve this issue, the AIA included something that didn't exist before—a technical definition of *inventor* in the law: "The term 'inventor' means the individual or, if a joint invention, the individuals collectively who invented or discovered the subject matter of the invention." Notice that it doesn't say you have be first, only that you have to invent. This definition, together with the derivation proceeding, appears to bring the AIA into conformance with Article I of the United States Constitution.

Secrecy and the European Model

If you came to this chapter looking for guidance on secrecy, you may wonder why you should wade through all the information about AIA, FITF, and derivation. The answer is that everything written about invention secrecy in the United States in all books published before 2012 is probably now incorrect and will probably lead you astray.

Patents Expert Pointer

Secrecy is critical before you file for a patent. Under the old US laws, you could tell everyone about your invention as long as you could prove you were the first to think of it. This is no longer true. It is best to follow the European model and keep in mind that one slip of the tongue (you tell one of the other dads at your kid's soccer game) and your invention could be rendered forever unpatentable. Wait until after you file before you tell anyone. The only exceptions are your attorney, your coinventors, your employer (if applicable), and anyone who signs a confidentiality agreement.

There may now be ten years of litigation on various cases before anyone really knows the impact of the new law on secrecy as the courts and the patent office struggle to apply the new rules to various real-world conflicts. The best guidance is to follow the European model if at all possible. For one thing, this helps assure that you can at least get a European patent, no matter what happens in the United States.

For the most important inventions, you are not going to want to choose to give up a market of 400 million Europeans and equally as many other possible consumers in additional countries that are aligning their patent systems with the European model. The European rule is simple: you can't tell anyone anything before you file.

In European patent law, there was never a grace period, and there was never any sort of derivation procedure to protect inventors from someone taking advantage of ideas they let slip. If you keep your idea perfectly secret, no one can derive their invention from yours. If someone else files your idea first, that is truly the end of the story.

Worse, in Europe even if you file first, another party could get your patent invalidated if they proved that you made your information public—whether or not someone else copied it. You cannot give a talk at a scientific meeting, and then decide to file a patent that afternoon. Once you have given a talk or published an abstract describing the invention, it is all over, for good. Neither you nor anyone else can ever get a patent on that invention. In Europe, there is no kind of litigation or court proceeding or hearing that can rescue you and your idea. It is over.

This harsh position is based on the idea that no one should be able to patent something that is already publicly known. Even if the inventor was the one who revealed it and is the same person who wants to patent it.

We can all see that it wouldn't be fair to allow a person sitting in the audience to file a patent on the invention revealed by the lecturer. In the European system, however, once secrecy is breached, it is considered that the information is available to all, and no one gets to patent it.

This is more similar to US trade secret law. Once a trade secret leaks out, it can't become a trade secret again. From the point of view of the patent system, the European concept is that the public has to learn of the invention from the patent document itself and not from any other source.

This makes a little more sense when you consider that fundamental bargain that underlies all patents: you agree to disclose all the details to the public if the government will give you a monopoly for a period of time. If you have already made it public, why should the government offer you a monopoly for something you have already done? In the common law of contract in use in the United States and England, this is called lack of consideration—the government shouldn't reward you to do something you have already done.

If you gave the lecture, but never filed the patent, the public would still have the information on how to practice your invention. What is the public getting in addition if it gives you a monopoly for twenty years? Seemingly, the public is getting nothing it doesn't already have.

Solving the Secrecy Problem: The Provisional Application

The best solution to the secrecy issue is to use the provisional application process. The patent system in the United States and Europe provides an opportunity for an inventor to file an informal preliminary patent application called a provisional application. This can be inexpensive and doesn't require hiring a patent agent or patent attorney.

With a provisional application, you can establish a filing date and prove you are first without having to come up with all the funds necessary for a full patent application. Once a provisional application is filed, you have one year to upgrade it into a full formal patent application. If you don't proceed to the full application, you lose the priority date, but you still haven't made the idea public.

If you have filed every detail of the invention in the provisional application, then you can freely discuss and publish this information without harming your ability to use all the information in the provisional when you do apply for a patent.

Use of Confidentiality Notices

Before you even decide about filing a provisional application, the best option to protect your ability to file is to simply make clear that the information you have is confidential. You may have seen or heard about complex confidentiality agreements, but you'll be happy to learn that you may get adequate protection by simply putting the word "confidential" on the bottom of every written page about your invention.

Use of the confidential notation won't solve all problems. It must appear that you really were trying to keep everything secret and limit the spread of information to a very few select individuals with whom you had to communicate. This might include going to your academic department chairman or to your boss to explain what you have done in order to get his support for the next step in the patent filing process. It might include a discussion with an engineer with whom you need to work to test the concept to make sure it will work as you think it will.

The essence of this idea is to provide a clear warning that the information is confidential, should not be released to the public, and is the intended content of an imminent patent application. When you make this clear on every page, a US or European court will probably accept that you have not violated the requirement for secrecy. This is because there is a clean, sharp difference between making information available to the public at large as opposed to making clear to anyone you communicate with that the information is confidential and no further release of the information is authorized.

If someone receives such well-marked information from you but violates the undertaking of confidentiality and releases the information, you will probably still be able to proceed with a patent application as if no release had taken place. It is always best to have some sort of simple agreement indicating that the person who receives or sees the information understands that you intend that it be kept confidential and that person may not release the information to anyone else without your permission.

There are limits to this. You can't get up in front of a thousand people at a public meeting, warn them that what you are about to say is confidential, and then give a lecture on the subject. But there is no hard-and-fast rule on when you have gone too far.

The best guidance is to discuss your idea with no one. If you must talk to a small number of individuals, you should communicate the information in writing so anyone can see exactly what it is you did or did not disclose, and every page should be marked at least with the word "confidential." You should also have a simple agreement that says the person agrees to view the confidential information and accepts that it is confidential. Then, as quickly as possible, get all the information into a provisional application.

T. A. EDISON.
ELECTRIC LAMP.

No. 251,546.

Patented Dec. 27, 1881.

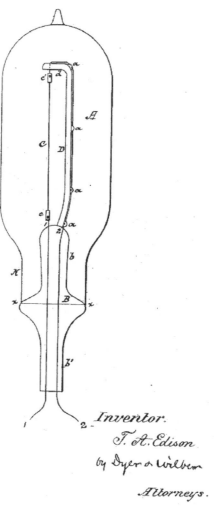

Witnesses:

Q.D. Mott

James A. Payne

Inventor.

T. A. Edison

by Dyer & Wilber

Attorneys.

Taking the First Step: The Patent Disclosure

"The specification shall contain a written description of the invention, and of the manner and process of making and using it, in such full, clear, concise, and exact terms as to enable any person skilled in the art to which it pertains, or with which it is most nearly connected, to make and use the same, and shall set forth the best mode contemplated by the inventor of carrying out his invention."

These are the words of a key section of the US patent law regarding filing a patent. It makes very clear that you have to fully tell how the invention is practiced. You, the inventor, can probably do this best, and you can do it initially as part of a provisional patent application.

What Is a Provisional Patent Application?

The most powerful tool available for the solo, academic, self-employed, or unemployed inventor is the provisional patent application. You can have an idea in the morning, and have a provisional application on file later the same day. The 2011 AIA has created a new category of inventor called the micro entity, and if you qualify for this, then the fee to file the provisional application is under $100.

Patent Vocab

A provisional patent application is useful for the solo or first-time inventor. It is a separate application that fully describes an invention and is a legally definitive statement you can write yourself. The provisional lets you immediately and inexpensively document what you have invented and the official date of your invention. The patent office keeps the provisional application secret. If you don't turn it into a regular patent application within twelve months, however, it more or less evaporates.

Filing the Provisional Patent Application

When I lived in London in the early 1990s, I would take the underground into central London, walk to the patent office, and hand in my provisional application with a cover sheet and £25. The clerk would stamp my application with the time and date, and that was it. Now, this can be done online at the USPTO website. You download the forms, fill in the blanks, add your invention description text, upload, pay online, and you are done.

It is instant, and it gives you twelve months to decide what to do next. It covers you for Europe and for nearly anywhere else in the world. Once it is filed, you don't have to maintain any secrecy to protect your right to file a patent.

The USPTO assigns a number to the document, but that is all it will do with it. There will be no examination of the provisional and no investigation of possible prior art by the patent office. It is filed away and can be produced at any time in the future.

What happens if the day after you file, you get a new idea that improves on the original concept? Well, you will have to pay the fee again, but you can instantly file that new provisional application as well. You can refer to the first provisional by number when you send in the second one.

Typically you will want to file a complete formal patent application within twelve months after you filed the first provisional. The patent application will mention the numbers and dates of the provisional and claim priority from the date (or dates) of the provisional(s).

The filing date of your patent is the date the actual formal patent application was filed. However, the provisional establishes your priority over any other inventor who comes in to file a patent on the same invention after the date of your provisional. Even if the other inventor has a full formal application written by attorneys for a huge corporation, your provisional will win the priority battle.

When Should You File the Formal Application?

The best time to file the full formal patent application depends on whether the filing fee (verify the exact amount with USPTO as fees change from time to time) seems expensive to you. You can file a provisional when you first have the flash of an idea, but all you will be able to protect is the basic idea as it occurred to you on that date. Then you can file a series of provisionals as the work progresses.

Ideally, the best time to file the formal application is when you actually get the invention to function because there may be some additional twist or adjustment or further step that is needed to make the invention have the utility, or use, that you intend. If you file a provisional with a description of something that doesn't quite work, and a competitor files a similar provisional with a description that has the extra few bits so that it actually does work, you will forfeit the patent. This is because even with priority for conception, the fact that your invention isn't working yet may mean that it won't ever work. You could get a patent on what you put into the provisional, but it would be a useless patent.

Patents Expert Pointer

Even though you can get a patent on an invention you haven't yet built and operated, it is a bad idea to do this. Once you have your invention working, then you are able to state all of the critical steps. If you get a patent on an invention idea that is missing a critical step because you hadn't realized you needed that step at the time you had filed, the whole patent could later be invalidated. This happens when you drag an infringer into court, who points out your description is missing a key step.

Let's suppose that the patent office ends up granting a patent to both you and your competitor. Any businesses that want to license a patent for this invention will license your competitor's patent simply because it protects an actual working invention. If you try to sue for patent infringement due to overlap with the similar aspects of your invention, the court will probably rule your patent invalid specifically because it does not work. The legal terminology here is *enablement*.

If you have a fully enabled invention, your patent could cover many different variations and ways of building and using your invention that you have not yet actually tried or tested. As long as one of the variations actually works, there is a good chance all the variations will be covered as well. If nothing in your description actually works, then you can't get usable coverage for anything.

Describing Your Invention

The moment you get your invention working, you will be able to write out a detailed description of it. The description of how it's built and how it works must be accurate enough that anyone who follows your instructions exactly will find that the invention works just as you say it does.

The description of the invention has to be detailed and it has to be sufficient. If you are using technical words in your description or any important words that have multiple meanings, then it is critical to provide exact definitions of what you mean when you use each of the ambiguous words.

Use clear, simple language when you're writing your invention description and reread it many times to find any discrepancies. Try to imagine how someone could read your words and misunderstand what you are describing, then rewrite any confusing passages.

Keep in mind that you are not writing for a complete novice; you are writing for someone else in a relevant field of work who is like yourself, but is considered ordinary and uninventive. If you are a maker of metal carving lathes describing a new type of lathe, then other lathe makers should be able to read your patent description and know immediately and completely what they must do to make the new kind of lathe you are describing in your application.

If you accidentally leave something out of the description that is critical, there may be nothing you can do about it and the patent will not be granted or it will prove useless because it does not reveal how to make the invention work, which is another way to fail on enablement.

A Provisional vs a Non-Provisional Application

If a provisional application has to have a precise and effective description, what makes it provisional?

The biggest difference between a provisional application and a formal, definitive application is that the formal application has to have at least one formal claim.

Claims

A claim is a precise legal statement of what is being protected. In other words, a claim restates the invention description in formal terms. Claims are exceedingly difficult to write, and there probably is no way you can safely and reasonably write a claim if you are not an attorney or patent agent who has passed the patent bar exam.

Since a claim doesn't describe anything other than what you've written in your description, you can't make claims of anything that goes beyond what you have described. Although claims are the heart and soul of the formal patent application, there is no need to include claims in the provisional application.

Inventor Beware

It is great to be able to file your own provisional application with an excellent description of your invention, but you will not be able to write the claims without the help of a patent attorney or patent agent. The main part of converting a provisional application into a full formal application is the addition of patent claims. Unlike the invention description in the provisional, claims are technical, and their wording depends on some very complex sets of patent law rules.

Seeking Professional Help

Other than that difference, however, the closer your provisional text is to the formal text of the description portion of the formal application, the better off you are. The best method to protect yourself is to file the best possible description of what you have done in the form of a provisional application, and then immediately seek the professional help of a patent agent or patent attorney to help you rewrite and refile the provisional as quickly as possible.

The patent agent or patent attorney knows a lot more than you do about what needs to be put into the description and how to adjust and optimize the wording to get the best possible protection. For instance, you might have written that "the new lathe has a novel secondary blade that is 1 centimeter in length." The patent professional might change the wording to, "the new lathe has one or more additional blades that are each between 5 millimeters and 20 millimeters in length, but preferably just one additional blade that is 1 centimeter in length."

You can see that your basic discovery is covered, but the expanded wording covers a range of variation that you might as well claim in addition. Your original description becomes the "best mode" that you, the inventor, knew of at the time of the filing.

Under the AIA, this best mode requirement is not as critical as it was in the past, but it is still a necessity. There are numerous issues like this that a patent agent or attorney knows about that can help bulletproof your provisional application and expand the scope of the formal patent application beyond what you had imagined at the outset.

The Pros and Cons of Filing a Provisional Application

The big advantage of filing a provisional application is that it buys you time to figure out if you want to spend the big bucks for a formal patent filing. It gives you time to do an extensive search of all possible literature to make sure no one has already invented the same device. It gives you time to develop refinements and additional aspects to include in the formal application. You can investigate the market potential to get a better idea of whether your invention will be worthwhile even if you do get the patent granted.

You can also find out more about ownership of your invention. As long as you are the inventor, you are able to file the provisional application in your own name, even if your employer ends up being the owner of the invention. When you first get your invention working, you may not be sure who can claim ownership.

Let's say you're a professor of microbiology at a major university and you remember signing an agreement giving the university ownership of any invention that results from your employment. However, you have just invented a better vacuum cleaner while working in your spare time at home on the weekend. Go ahead and file the provisional application, and then find out whether you or the university owns it.

The formal application could cost anywhere from a few thousand dollars on up to a few hundred thousand dollars to carry it on through to grant of the patent. You are going to need to line up funding either by setting up a company and getting investors, identifying an industrial partner willing to work with your university or corporation, remortgaging your house, or simply making a series of presentations within your university or corporation.

This is where the concerns about first to file vs. first to invent really have an effect (see Chapter 6). In a big corporation that files hundreds of patents a year, the entire process is very fast, so if you and a corporate inventor independently have the same idea at the same time, the corporation will rapidly have a formal application filed while you are still trying to get some investors to meet you for lunch. It will be first to file and you will get nothing.

In the old US system before the AIA, you just needed to document that you had made the invention by writing it all down in a lab notebook and having someone witness and sign it. As long as you kept working on it, you could spend three or four years lining up investors. Starting in 2013, however, the rules will change.

Under the AIA, the only action you can take to reliably preserve your rights before filing a formal application is to file a provisional application. Only a provisional application filed by you before another inventor files either a provisional or formal application for a similar invention is going to reliably win the race for you.

AIA: Change of Rule

For over two hundred years in the United States, inventors documented their ideas by writing them down and getting a witness to sign and date it. The AIA has made that approach useless. Starting in 2013, the provisional application is the only way you can get preliminary proof of your invention.

The Cons of Filing a Provisional Application

So what is the downside of filing a provisional application? Mostly, it is lost time in the start date of enforceability. If you are in the pharmaceutical industry, for instance, and it will take twelve years to bring your product to market, then the filing date or date of grant of the patent won't matter very much. By the time your new pharmaceutical goes on sale, the patent will have been granted for years, so it doesn't pay to file a provisional application.

By contrast, let's say your invention pertains to a new app for the iPhone. Even if your formal application is filed tomorrow and gets rocketed through the patent office to grant, it will take many months, even a couple of years. However, if you're in this type of industry, you will want to be able to enforce the patent against infringers as quickly as possible. You won't want to let an extra year go by before you file your actual application.

Also, not having claims could be a liability. The process of preparing the formal application, doing the prior art searches, and answering the examiner's questions can result in very important improvements in the strength of your patent. By postponing these activities, you could create an opening for a competitor to make off with some of the rights you could have had for yourself.

You could also make statements in your original draft that cause harm later on down the line that a patent professional could have helped you avoid. Fortunately, the provisionals all remain secret and get destroyed eventually if they are never cited by a corresponding formal patent application. You can redraft your application and just lose a little bit of time on your priority date.

For the most part, the positives of filing a provisional patent application outweigh the negatives. The provisional process gives the noncorporate inventor a chance to balance things out vis-à-vis the competing inventor in a large corporation. It provides excellent proof of exactly what you knew and when you knew it. It even allows you to start stamping *Patent Pending* on any products you actually make and sell while you are deciding what to do.

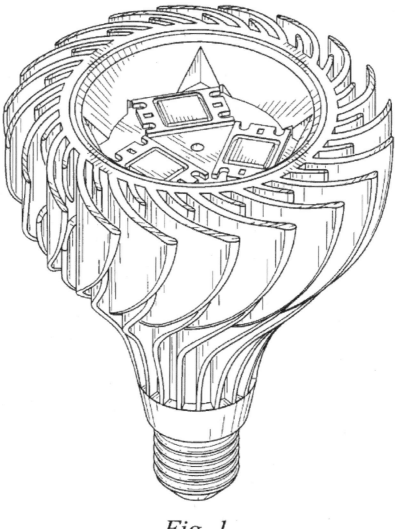

Fig. 1

CHAPTER 8

 **Prior Art**

> ## In This Chapter
>
> ➤ What the patent office considers new
>
> ➤ Researching previous similar inventions
>
> ➤ The prior art search

Determining the existence of prior art is a most important step when preparing to file a patent. Because of this importance, many tools are available to help you search for prior art related to your invention. Let's take a look at prior art and what it means to you as an inventor.

What Is Prior Art and Why Is It Important?

One of the fundamental rules of invention is that you cannot call something an invention if someone else has already thought of it. *Prior art* is the term used in patent law to describe any relevant knowledge in the field of an invention that existed prior to the moment of that invention.

What if you showed that a vast amount of prior art exists, but none of it describes or predicts your specific invention? Then the prior art proves helpful because it shows that your advance is nonobvious—many people were working in your area and much was known without anyone realizing what you have discovered.

You Can't Be Second to Invent

If you find evidence that someone previously invented exactly what you have just invented, then that prior art makes it impossible for you to get a patent. It is said in that case that the prior art anticipated your invention.

In some cases, the prior art comes near to what you have done but doesn't quite match it. Perhaps another person wrote a paper drawing attention to the fact that existing devices didn't solve an important known problem and made some suggestions about how the problem should be solved. Even though that person did not go the last mile to actually build the new device, the information and suggestions might be so close to what you have done that a court could determine that your invention is obvious in light of the prior art.

Patent Vocab

Prior art is a term the patent office uses to describe any information that is relevant to deciding whether your invention is truly new. It includes everything before your filing date such as:

➤ Patents

➤ Academic publications

➤ Magazine articles

➤ Posters

➤ Lectures at professional meetings

➤ Public demonstrations of a similar device

The term can be so broad that it could even include some kinds of information that has never been made public, as well as something you may have revealed publicly yourself before you filed.

The patent office or the court could put together two or three different papers that each hold a piece of the puzzle and say that since anyone could have read all three papers, your invention is obvious in the light of the combination of prior art references.

Let's say you invented something and then toiled away in secret, developing it for three years. Unknown to you, someone else makes a similar invention two years after you did, files a provisional, publishes, and then files a patent. When you discover this, you rush to file your own patent. What happens?

The patent office will just say that the competitor's invention is prior art to your filing, and you will be out of luck.

In the old US system of patent interferences and first to invent, you might have had a chance to prove that your work had priority even though the other inventor could not possibly have known what you were doing.

Under the 2011 system of the AIA, however, the person who is first to file gets the patent. Your only hope is a derivation proceeding (see Chapter 6), and that is applicable only if you can prove that the other inventor actually got the information from you.

Focusing again on your own situation, you might wonder if you are responsible for knowing everything. Think of how unfair this could be. Let's suppose a machinist in communist Hungary in 1957 figured out how to make the same device that you just invented. He wrote up a description in Hungarian and filed it at the local town hall in a small city 200 miles from Budapest. No one else ever heard about this. He died soon thereafter—with no descendants—before he even made the first one for sale. You knew nothing about this.

You have been granted your patent, you started a company, you have gone into production, and the invention is a huge success. A knockoff specialist sees your success and decides to copy your invention and try to overwhelm you in the marketplace with intensive marketing done by attractive sales associates. You sue for patent infringement. The specialist discovers the Hungarian write-up and seeks to have your patent invalidated based on prior art. Is there any way this nightmare scenario could really make the specialist the victors? Well, yes, it could. There is probably very little you could have done to prevent this disaster.

There are a few particular rules about this. For instance, if the town hall property clerk created an index of invention filings and published this annually in the county newspaper, you are almost certainly done for. If the Hungarian machinist's papers were stored away with no indexing, your opponent probably couldn't have discovered it, and even if he did, you might be able to succeed in fighting the invalidation.

The issue with central Europe is that Germany, Austria, and other countries have had a system of *Gebrauchmuster* filings on a local level that provided inexpensive local protection for area craftsmen. This may have gone on for centuries.

If the silversmith in a small town discovered something in the 1600s that improved his product, he could store a description in the town hall and get the local council to give him a privilege to use his invention and exclude the other silversmiths in town from copying what he had done. You can readily see how in another age local protection made sense.

With the elapse of time, the fall of communism, the progress of electronic media, and the globalization of the economies of the world, this sort of local prior art is fading away into insignificance as a threat.

It is not a question of whether it's fair to hold an inventor responsible for difficult-to-find prior art. It is not a question of being responsible for knowing about obscure prior art. The only thing the PTO, the court, and your opponents care about is whether relevant prior art exists, and, if so, is it sufficient to anticipate your invention or to render it obvious.

Prior Art Is Your Friend

One helpful mantra here is to keep telling yourself during this phase of planning your patent that prior art is your friend. Here's what you don't want to do: you don't want to have a psychological block or a fear of looking somewhere because you suspect there could be prior art hiding there. You may harbor the feeling that if you don't look, you won't have to know about it (the ostrich head in the sand approach). Maybe no one will ever look at this possible source. You don't want your dream to be dashed by inconvenient facts. Well, this is a really bad approach and is something you just have to get over—completely.

Patents Expert Pointer

When an inventor begins a search for prior art, there is a natural psychological tendency not to want to look very hard. This is because the inventor fears finding something that will ruin his new perception of being a freshly minted brilliant inventor. The patent expert knows this emotion and is able to completely overcome it. If you find some relevant prior art, you can rush back to the drawing board and invent further improvements that really are not anticipated by prior art. This gets you a more invincible patent from the get-go.

The last thing you want to do is give up your job, mortgage your house, turn your life upside down, abandon all hobbies and friends to make your invention a success, and then find out three years into development that it was totally anticipated by a piece of prior art. It is much better to be aware of prior art as soon as possible.

The benefit of discovering prior art before filing your patent is that it can actually make your invention and prospective patent stronger. If you find that something very close to your idea was invented, you can learn what was wrong with it. Why didn't it become a hit? Is there some additional step you can take? Look into what the inventor or craftsman is doing now. Focus your search. Run out all the leads. Get as current as you possibly can, then reframe the problem and invent well beyond the prior art if you can.

Being aware of prior art gives you two choices:

1. Abandon your dream of your own free will and not waste more time and money on an unpatentable invention.

2. Develop a stronger, more certain vision of what your invention must be in order to succeed.

How Do You Conduct a Prior Art Search?

Conducting a proper patent search for prior art used to be a specialized process. Recently, it has become possible for you to do nearly all your searching on your home computer with very little expense or effort.

What Should You Look For?

There is a list of four types of prior art that your search needs to ferret out:

1. Any prior patents

2. Any prior printed publication about the invention

3. Any public use or sale of the invention (in the United States)

4. Any reason the invention was known (in the United States)

Due to the AIA of 2011, all four of these types of prior art occurring anywhere in the world will affect the US inventor. Until now, the idea had been that a US inventor could readily access written material from anywhere in the world, but could be affected by public use or knowledge without a written record only if it occurred in this country. Starting in 2013, the inventor will be responsible for unwritten use or knowledge, as well as all written prior art, anywhere in the world.

AIA: Change of Rule

The new AIA rules reflect the fact that the whole world is now online. In the 1950s, you could not possibly know about inventions being presented at the town square of small towns in foreign countries. Today, there is probably a video of that event on You Tube—or a video of someone describing it—that you could find and view in an instant. You are now pretty much responsible for all prior art everywhere.

You are going to be most concerned, of course, with any prior art that directly anticipates the potential claimed parts of your invention. There are three different ways in which highly relevant prior art can affect your ability to proceed:

1. Any such prior art that was publicly available before you invented the device will probably put an end to your invention

2. Any anticipating prior art that was available more than one year prior to the date you file your patent application will probably put an end to your invention. This is true even if you were the one who made it public and the prior art is your own invention—prior art against yourself. This is also true if you filed a patent application in another country and later decided (after a year) to file in the United States. The only way you may get off the hook is if the prior use of your invention was part of necessary experimentation to develop the invention.

3. Anticipating prior art that was secret—not available to the public—can affect you if you fall under the old patent laws and your invention becomes part of an interference proceeding (see Chapter 6). Under the new patent laws, secret prior art will be a problem if it can be used to prove that you derived your invention from someone else's invention.

AIA: Change of Rule

Secret prior art is slightly less of a problem for inventors under the new AIA. It can still be used against you, but only if it was part of a patent application or a patent development project you might have learned about.

The AIA may have an effect in this area, but the fact still exists that an unpublished secret prior art reference can be used to invalidate your patent.

Basically, there are two types of secret prior art that can affect you:

1. A patent application or patent that has never been made publicly available. This is particularly common in patents with military uses.

2. A competitor who can prove he invented first and had not abandoned the invention, but was slowly developing it could invalidate your patent during a litigation.

The Search

You may be tempted to use the Google patents search engine to begin your search for prior art since the Google name is so famous for searching. This is not reliable, however, because Google patents doesn't yet offer a comprehensive list. I have a dozen granted patents, but when I search on Google patents for inventions under my name, several don't appear. Google keeps improving its patent search function, but the problems still present add up to one conclusion: don't even think about relying exclusively on Google.

The UPTO web site allows patent searches and Delphion has a good subscription web site. Delphion, the European Patent Office, and the World Intellectual Property Office can all help you track down not only granted patents, but also pending applications from around the world. In many cases, patent applications are not publicly available for eighteen months after they are filed. Any prepublication prior art hidden in the patent application system is just simply out of your reach.

In academic or scientific subject areas, you should do searches of the professional published literature on any topics that seem to be closely related to the subject of your invention. As you know, this will only show articles that have been published. You also need to know what is being worked on currently. You can search grant applications and grants with the major funding agencies. You can also search meeting proceedings that will have published abstracts of presentations given at scientific meetings, or sometimes just the title of talks in a meeting schedule or brochure is available online.

You should attend trade shows in your field to check what exhibitors are showing and to hear the most recent talks in your field. If you do this during the year or so following your patent application filing, you may get insight into what a particular researcher or group was doing since the previous meeting the year before.

The Implications of Your Search

At some point, you will believe that you have done the maximum prior art search possible. You might have missed something, but you have looked diligently. Could you have done more? Here is how I would think of it.

Put yourself in the mind frame of your worst enemy—the intentional competing infringer of your invention. You are suing the infringer for millions, and he will be bankrupt and lose everything when you win. His only hope of saving himself and ruining you is to search and search and search until he finds that one piece of prior art that can decimate you and your patent. Would that infringer be working harder than you have worked to try to ferret out any relevant prior art?

Be Honest

You must tell the patent office if you find anything relevant when doing your prior art search. If you don't, you are considered to be engaged in fraud. And if you keep the finding secret and the court or your competitors later discover that you knowingly withheld prior art, then your entire patent—all of the claims—will be rendered unenforceable even if you had one hundred claims and only one was affected by the covered-up prior art. This kind of fraud has other negative effects on your credibility in other areas, so don't even think of it.

Inventor Beware

You may be tempted to keep relevant prior art that could wipe out your inventions secret from the patent office. Definitely do not do this; this is fraud, and you are trying to get a government monopoly that you know you don't deserve. Even if you manage to keep secret the fact that you knew about the prior art, someone else may find the same prior art and invalidate your patent years from now thereby turning all your time, money, and energy spent into a colossal waste. Search diligently and tell the patent office what you find.

On the other hand, if you pass along a possible prior art reference that seems likely to knock out or seriously affect one or more of your most important claims, your patent will definitely take the hit, and the claims in question could be removed, but the rest of your patent will still be in the running.

Some patent attorneys have argued that it is better not to conduct a vigorous search for prior art. The argument is that prior art will only affect you if someone finds it. If you find something and don't think it's relevant, not reporting it could cause a lot of trouble if a judge later decides it was.

Keep in mind that the patent attorney probably gets paid whether the patent goes through or not, but you, the inventor, may have sunk everything you have into this invention. There should be no doubt on this issue. Do the most intensive prior art you can imagine, and do it early before you submit your application. Then repeat it once you have filed. You must report everything you find to the USPTO.

More Reasons for Searching

If you find the prior art before you file, you can amend and update your description to go around or beyond the prior art. If you find the prior art after filing, you can still make adjustments in your claims to work around the prior art.

Carefully reading and considering all the relevant prior art you can find will help you identify and clearly express the important ways in which your invention differs from what went before.

If someone seems to have come close to but falls short of your invention, think carefully about why. Did that person reach a wrong conclusion that led him away from the insight you had (this is called teaching away)?

For instance, you might find that three important leaders in your field recently concluded their published papers by saying that the future of the field depended on finding better ways to suppress a particular type of noise that was confusing the process of data collection. However, you discovered that the noise contained more important information; the ordinary data needs to be suppressed and the noise amplified to make the breakthrough.

The more aware you are of the history of your field, the better you will understand it. The more expert you are about what everyone else in the field has done, the better you will be at enhancing the description of your own invention and the better able you will be to defend your patent from attacks in the future.

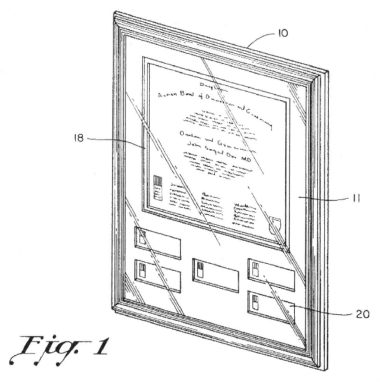

Fig. 1

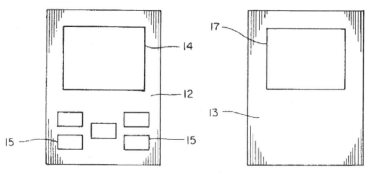

Fig. 2   *Fig. 3*

CHAPTER 9

Filing a Patent Application

In This Chapter

➤ The patent application

➤ Understanding your audience

➤ Insider tips

The biggest formal step toward getting a patent is drafting or writing the full patent application. This is a formal process and given the expense and legal challenges you could face in the future, you have to be sure that the application is the best possible expression of your invention.

Who Should Write Your Patent Application

Because of the formality of the writing and because of the complex structure of the claims, it is unlikely you will be able to write your own patent application without the help of a fully trained, expert patent agent or patent attorney.

A general attorney can't do this for you; few attorneys have enough knowledge and experience to help you if they haven't actually taken and passed the patent bar exam. Even an attorney who has been involved in patent litigation, representing a party in a patent infringement law suit, may not be sufficiently experienced in actually writing a patent and its claims.

Writing Your Own Preliminary Application

Writing your own preliminary version of the patent application, is a good idea for a couple reasons:

➤ Writing a description of your ideas will help you clarify and formulate what you are thinking. The patent application is a form of communication in which you tell the world what you have done and how anyone else can do the same once the patent expires.

➤ You may be able to control and reduce the expense of having a patent agent or attorney write the definitive application if you do as much as you can to get everything together in the right form with the correct components. The better you can describe the idea, the more your attorney or agent can focus on producing the best translation into a formal patent application. You focus on the fundamental description of the technology so the attorney or agent can focus on the legal side of the document.

Patents Expert Pointer

Even though you will definitely need the help of a patent attorney or agent to produce an excellent patent application, you can do a great deal yourself. This will help you make better use of the expensive time of the patent professional. It may also be an opportunity to discover more about your invention by trying to write about it. You can even try your hand at some basic claim writing, even though you must have every claim written by a professional before the application is submitted.

Many inventors can build a great invention but are not able to write about it. In that situation, they simply describe the invention verbally and answer questions until the patent specialist understands the invention enough to write the application.

If you have written a provisional application, you should allow about two months for a patent attorney to convert it into a formal application. For this reason, you really have about ten months after filing a provisional before you have to line up the funds to have a formal application written up. Getting a patent written, filed, defended, and granted could cost as little as $5,000 and as much as $500,000, depending on the complexity of the invention.

In the appendix I have attached one of the simplest, least expensive patents I've filed, as well as the most expensive, elaborate patent. The large, complex one (US 5,560,360) may seem like overkill, but it has survived numerous intensive counterattacks during several patent infringement litigations over a number of years against the most well-funded adversaries in the world, and against some of the world's largest law firms.

Patent law changes over time, but the patent you file today will have to hold up against changed patent laws fifteen or twenty years in the future. This type of durability against future changes in patent law may seem impossible to achieve, but it is also the hallmark of an excellent patent application.

Components of a Patent Application

A patent application has several specific parts that have to be filed in a specific order. The patent office provides a list of the order of parts as it likes to receive them. The organization in this chapter is more directed to how you should think and write.

Inventors and Oath

It may sound way too obvious, but you must first correctly identify who the inventor or inventors are. If you worked entirely on your own, then this is indeed very simple. You are the inventor and you are the only inventor.

The process is different than what is used to determine authorship for academic papers. If you are in a big laboratory at a university and you invent something, the director of the laboratory is not an inventor, the principal investigator of your grant is not an inventor, your coworkers who ran the experiments with you are not inventors, the associate who did a confidential, careful reading of your first draft of the application and made some suggestions on wording, clarity, and structure is not an inventor. Only persons who actually contributed significant inventive ideas to the conception of the invention are inventors.

If you sue someone who infringes your patent, the infringer may be able to derail the litigation if he can prove that you included people as coinventors who did not really invent or if you omitted people who really did make a significant inventive contribution. Be sure that you get this right.

If you listed people in your application who did not invent and the reason they were listed is innocent—a simple mistake or misunderstanding—they can just be removed from the patent application without much trouble.

But if there is evidence that names were added for reasons that affected the filing date or any other sort of intentional manipulation of the truth, then serious trouble can ensue. And if you omitted someone, the trouble gets much worse because you do not have an exclusive right to the patent since each inventor is entitled to use or assign his rights. An omitted inventor who did not assign his rights along with you can disrupt your standing, or right to sue, or can provide an opportunity for an infringer who may be able to get a license from the neglected inventor.

Each true inventor must sign an oath that has serious implications. The most important components of the oath are:

➤ A statement that the inventor believes that he is the true first inventor of the invention

➤ That the inventor has disclosed any prior art that the inventor is aware of

The Title

There is no special requirement for the title. This may seem like a trivial issue, but there are important matters to keep in mind. In some industries, competitors try to hide their patents in plain sight: they choose titles that say absolutely nothing useful or relevant about the invention.

This is common in the pharmaceutical industry. As an example, two of my patents were drafted and filed on my behalf by a large pharmaceutical firm. It chose the title of the first patent to be "Particulates" and the title of the second patent to be "Particulate Agents" even though both patents related to a very complex advance in neuroscience.

Giving a patent an ambiguous title made more sense in the old days of patent searching, when many search methods relied on scanning the titles of patents. The downside is that this may lead competitors to infringe on your patent simply because they failed to discover it.

Most patents are given formal descriptive names that tell a great deal about the key point of the content.

The Abstract

The abstract describes the essence of your invention in 250 words, and it appears at the beginning of the patent. But writing the abstract before you write anything else is a great exercise to get you to focus on several key points:

➤ The field or subject matter of the invention

➤ What the invention accomplishes

➤ The key unique feature of the invention

A brief example of an abstract could be as follows:

This is an invention in the field of technologically advanced farm plows. It achieves improved crop yields by using a global positioning system to control the structure of planting rows and seed density. Additionally, it is new in that it is equipped with an ultrasound and laser rock detection system to allow full robotic plowing. The novel detector can allow the plow to lift a plow segment over any rock or push the rock away with a robotic arm.

As you can see, the abstract lets a person searching the patent know what kind of technologies may be involved. It also notifies the patent office about the fields it will need to search in, the experience it should look for in selecting the best examiner, and the scope of the key novel aspects that should get the most attention.

The Background

The background section is a detailed explanation for the patent examiner of the current state of the art in your field and explains why your invention is needed. It should explain the shortcomings of existing technologies so that the examiner understands the new use you are proposing. The new use is to solve the problem that other technologies have failed to do. You might also explain why existing technologies have not worked.

The Summary

Following the background is a summary of the invention that goes into greater detail than the abstract. Here you describe in overview all of the components of the invention and how they work together to solve the problem posed in the background section. The main purpose of the invention should be clear, as well as any additional purposes or uses it may have.

An example of a summary for the high-tech plow could be as follows:

The purpose of the invention is to solve the last few problems that have prevented fully robotic plowing without a human presence. In addition, the invention reduces the rate at which the plow blade needs to be replaced and allows for lighter construction because rocks are removed instead of being casually hurled up at the undercarriage. With the increase of crop prices, land that was not used for the past century is now being turned into farmland, and this robotic rock clearing system will greatly reduce the cost of plowing new working land.

Brief Description of the Drawings

Drawings of a new invention often help the patent examiner and other readers to understand what you are describing. Until 1880, the USPTO required the inventor to provide a working model of the invention with most applications. Since that time, drawings alone have done the job.

Patent drawings have a distinctive recognizable appearance because of the specific rules the patent office provides that detail how to do a drawing for a patent. There are typically different views of the invention, including cutaway views, enlarged views, and as many drawing parts as are needed to show every part that is claimed. All of the rules are available for download from the patent office.

Instead of labeling with words, patent drawings usually have a series of sequential numerical labels so that each component can be referred to by number in the patent description.

Overall, the style and shading and size of the drawings are based on the type of printing the patent office used for many years to publish patents.

It is possible to petition the patent office to allow you to submit color drawings or photographs, but you have to make a compelling case. Issued patents, as of the time of this writing, do not include color and do not adequately reproduce photographs. A notice is included in a patent that has color drawings or photographs, and any interested person must then request that a special copy be made and pay a fee to cover the costs of reproduction.

In addition to drawing out the shape of a mechanical invention, the drawing section is used to lay out complex block diagrams that may show all the logical portions of the invention. This is especially important in software inventions, for instance. Each block in one diagram may be broken down into subblocks in another diagram. There can be loops and logical branches drawn out in this fashion. Each block can contain some writing, but the blocks themselves are numbered for reference in the description.

The written specification or description includes a brief explanation of each drawing, which further identifies each labeled item or block in each drawing. The description should briefly point out how the given drawing relates to the overall invention.

Description of Preferred Embodiment

The greatest level of detail goes into the preferred embodiment, which is a description of how the invention is made and how it can be used. There may be several preferred embodiments, with at least one being called the best mode.

The description portion of the specification has three important tasks that are outlined in patent law:

1. It should provide a full and detailed description of the invention. It must be sufficiently clear and precise that any ordinary person who is skilled in the area of the invention would be able to make or use the invention.

2. The specification has to make absolutely clear which parts of the invention are new and which are old or existing parts. This must leave no doubt about exactly which aspect or aspects of the invention are going to be the subject of the patent.

In the case of the high-tech plow, it may turn out that putting a GPS into a plow to make it robotic is well known. However, previous systems have used laser or infrared systems to detect rocks, but the new system is unique because it mixes and compares ultrasound and laser data to improve the rock detection process. Every other aspect could be part of the prior art.

In this example, you could just describe a laser-ultrasound system, but it might turn out that this is widely used in neurosurgery for brain tumor identification, for example. The reason it is an invention is that it never occurred to anyone that such a system could be useful in farm plows. If you only described the electronics of the laser-ultrasound comparator, it might not be novel.

It is true that if your invention was a new type of laser-ultrasound comparator, then you could just describe this device and mention that it can be used in both brain surgery and farm plowing. However, in this particular invention, you are showing how a new machine, a farm plow informed by the laser-ultrasound system, achieves a valuable improvement in utility.

You would not need to spend much time describing the remainder of the plow system (the blade, the engine, the wheels, the drive shaft, the computer) but would focus mainly on how the laser-ultrasound system was built and operated, and how it interacted with the relevant components of the plow and tractor systems.

3. The specification must describe a particular fully working preferred embodiment, or best mode.

There might be fifty different ways to build, assemble, and operate a device that falls under your inventive concept. They are all different ways of achieving the same objective but share in common the unique part of your invention. In our plow example, it might be possible to mount the detector on the top of the vehicle, on a boom held out in front, on the undercarriage, or at the location of the headlights. The boom-mounted version could be an overhead boom or a pair of booms on either side. The booms could be movable or fixed, and so on. All of these would be covered, but you need to pick the particular arrangement that you, the inventor, believe works best based on your experience and opinion at the time the patent application is filed.

You need absolutely every detail of your invention's construction fully laid out in the patent. For a chemical patent, this could be a detailed laboratory method, including the weighing out of so many grams of one compound or another, spinning in a centrifuge for an exact number of minutes and at an exact number of gravities.

If you omit a critical step in the method, something that requires an ordinary person to figure out what to do, or that requires the creativity of an inventor, then the disclosure will not be sufficient. This can lead to the entire patent being invalidated.

The Best Mode Requirement

The AIA of 2011 loosens the requirements for the inventor to describe the best method, or mode, for using or building the invention. The preferred embodiment must still be described in great detail, though, to avoid the possibility that an inventor, wishing to get a twenty-year monopoly in exchange for disclosing the invention, will provide an incomplete disclosure.

An inventor who knows that the method he wrote out in the patent worked, but just barely, and that there is an alternative means of practicing the invention that is faster, more reliable, and produced a far better product is breaking the fundamental bargain of the patent by not disclosing the best method.

AIA: Change of Rule

There may be a variety of different versions or ways of practicing an invention. US patent law has always required that the inventor describe what he considers to be the very best of these different versions when the patent application is written. Under the AIA, however, an invention can't get invalidated in court on this matter unless it can be proven that not describing the best mode was intentional. It is still safest to describe the very best method you can think of when your write your patent application.

Congress left the best mode requirement in the law, but ended its use in patent litigation mainly because there is no best mode requirement in the European or Japanese patent systems. Patents arising from inventors outside the United States often would not meet the best mode requirement and would get invalidated in the US court while being valid elsewhere. So in the interest of global harmonization, there was pressure for congress to weaken or remove this part of the patent law.

The best mode requirement was despised by overseas companies and individual inventors especially because it led to expensive litigation.

Consider what is involved in an attack on a patent in court based entirely on failure to disclose the best mode. The attacking party has to take depositions and carry out extensive discovery—poring through every note, e-mail, experiment, and interoffice discussion that occurred ten or fifteen years in the past—to try to prove that there was another way to perform the invention, that it was better, and that the inventor intentionally didn't disclose it.

You might think that such an attack would fail because it is so difficult to find this sort of information, but the actual effect would be to cause litigation costs and time to spiral and grow. Further, a judge would have to decide if a particular method would be better than another based on the quality of the argument made by an attorney, and this often appears to lead to unfair invalidations.

Under the AIA, the best mode is still required, but it affects the quality of the original disclosure when it is examined by the USPTO, rather than being a point on which the patent can be invalidated later in court. It is still possible that an intentional failure to disclose the best mode will be treated as fraud on the patent office, so a patent could be rendered unenforceable in a patent litigation if you are caught doing this.

This scenario differs from the previous situation because it usually requires an intentional failure to disclose best mode. Under previous law, the simple fact that a better mode existed, even if this was only understood in hindsight years later, could be used to invalidate the patent.

The bottom line is that for at least three reasons, it still makes sense to disclose the best mode.

1. If you have a business, you are going to want to use the best mode so that you can make and sell the best possible product.

2. You want to make sure the patent protects every aspect of your best mode method, so it is best to have it in the patent.

3. If you keep the key information out and hold it as a trade secret, any leak of the information will give you one shot to litigate—for misappropriation of trade secret against the first violator—but then everyone else can use the information freely. With a patent you can stop everyone who tries to use your method without permission.

Reduction to Practice: Actual vs Constructive

A helpful aspect of the description is for it to prove beyond a shadow of doubt that the invention was up and running when the patent application was filed. This is called reduction

Patent Vocab

Constructive reduction to practice occurs when a patent application is filed with the USPTO. Any average person should be able to use the invention without conducting further study.

Actual Reduction to Practice is a physical representation of the operating invention or invented process.

to practice, meaning that the thing really works.

You can imagine a situation involving a complex mechanical patent in which an inventor has had a great idea about a new type of machine that will do some great new process. He is excited about the invention, believes it will be incredibly valuable, and that numerous competitors will want to steal his idea. Granted, there may be a bit of unsupportable paranoia in this fear, but it is a common emotional state for inventors.

Because of the emotional aspect, the inventor is getting his patent application ready to file even as he is struggling in the lab to get the machine to actually work as he envisions it. He thinks everything is nearly working; all the problems he knows about seem to be solved. He is very far along in constructing the first working model and his fears are growing. He tells the patent attorney to pull the trigger by filing the patent to get the protection in force. However, the inventor's lab work hits a snag just after the filing. Something he didn't think of or know about turns up as a huge problem. For instance, a critical step results in explosive fracturing of the most sensitive part of the machine every time it is actually turned on. Later, the inventor solves the problem and starts producing the machine and has huge success.

The patent goes forward and get's granted. Years later when the inventor tries to stop an infringer, the infringer follows the description, builds it for the court, and shows that it explodes and fails every time. This reflects an inadequate description, and the patent will probably get invalidated.

If you have actual reduction to practice, you know your invention works, you can be sure that your detailed description in the patent application is sufficient.

When you submit a patent application, you have constructive reduction to practice, which means that the court assumes the invention is working based on the fact that you have filed. However, the best way to bombproof your description is to hold off on the patent filing until you know it works in the real world and that you have exactly described what is needed to make it work.

With the old first to invent system, you could just keep on making new entries in your lab notebook and get someone to witness and sign it as each new step came along. This would provide instant protection.

Under the new AIA of 2011, however, the lab notebook system does not work. If someone else independently comes up with the same invention you have and files before you do, all of your lab notes proving your invention predates your competitor's by two years are now useless. The patent office and the courts no longer care. This new situation is said to encourage a "rush to the patent office." Don't get burned by rushing in before you have actual reduction to practice and a completely reliable description.

Claims

The claims are a key part of the patent application. They define the exact invention from a legal point of view. Every formal patent application must have at least one claim. Most applications have numerous claims—twenty or thirty claims are common. You will rarely see as many as one hundred, although there is no limit. The patent office just charges more per claim, and your attorney will probably charge more as the number of claims go up.

The language of patent claims is technical, so it is nearly impossible for inventors to write their own patent claims effectively unless the inventor is a patent agent or patent attorney who has passed the patent bar exam. That should not stop an inventor from writing out a simple version of some of the claims to help communicate with the patent professional.

To do this, you try to write out a single sentence that captures the key elements of your invention. Different claims arise for different aspects of the invention. You might say in our plow example:

1. I claim an automated plow system that relies on GPS for guidance that is equipped with a laser/ultrasound system for rock detection.
2. I claim the system of claim 1 that has a robotic arm that can pick up rocks.
3. I claim the system of claim 1 that has movable plow segments that can dodge detected rocks

Chapter 11 is devoted to details about claim writing. At this point, the most important message is that you need to be prepared to struggle to understand the claims of your patent no matter how technical they may sound. If you don't understand the claims, then it is possible that the patent professional has not fully understood the invention and has left something out or misstated something.

Appendices

One part of a patent application that often doesn't get mentioned is the appendices. Most patents do not have any. This is another trap for inventors. Long ago, the patent office halted the process of requiring models of inventions, but the courts may still want to see a surprising amount of material when your infringer's attorney starts telling the judge how much he thinks you have left out.

For software patents, there is no harm in attaching a full printout of the software as an appendix. If you don't do this, here is what can happen:

You say in the application that you have an algorithm in your word processing program that fixes presumed incorrect capitalization when two capital letters follow each other in an otherwise lowercase word as the person types. You have a box in the algorithm section of your drawings labeled Capital Letter Correction Algorithm. In the application, you describe the critical steps of the main program.

Your invention can get invalidated if your opponent argues that there is not enough information to allow a noncreative programmer to write a word processing program with that algorithm. You beat this by attaching an appendix with the entire code. This does mean revealing more than you have to, but it also means that your patent is bulletproof so you can stop someone from infringing your patent.

Similarly, if a critical aspect of your invention uses a mathematical algorithm or a seemingly well-known chemical process in a classic published article, you may want to attach the article or at least the relevant section as part of an appendix, which would be stored in microfiche form at present and then referenced in the application.

The concept that all the information you rely on and that another person needs to duplicate your invention should be present and specified is called intrinsic evidence, or the four corners of the patent application. The judge or the competitor should not have to go looking elsewhere for extrinsic evidence to find out what you mean.

Concepts to Keep in Mind

Here are some helpful hints to keep in mind when writing your patent application:

> ➤ Write with the appropriate level of detail
>
> ➤ Make sure your words mean what you think they mean
>
> ➤ Understand who your audience will be

Revealing Too Much vs Revealing Too Little

There are competing schools of thought about how much detail to reveal when writing a patent application. My own position is that you should reveal everything you possibly can. Many experienced patent attorneys, however, advise the very careful use of words so that you can get the protection you desire without giving too much information away to potential infringers.

The risk with being too selective is that you will leave out something that later may prove to be critical if the patent ever ends up being used in patent infringement litigation.

When you leave information out of your application, you effectively create trade secrets around the margins of your patented invention. If your patent is never used in litigation, then this could be a reasonable strategy. The patent warns competitors to stay away, but it doesn't give them enough information to be able to compete effectively. At a variety of levels, holding back is not consistent with the fundamental bargain of the patent, which is to disclose everything in exchange for a monopoly.

This type of secretive gamesmanship is most common in large corporations with a large number of patents, particularly in fields where infringement is common. For most inventors reading this book, I would advise not to consider using this strategy.

Definition of Terms: Being Your Own Lexicographer

The vast majority of patents have a fatal flaw: they lack adequate definitions of terms. This is actually bizarre—to the point where it seems to be a conspiracy of patent attorneys against inventors. If you look at most commercial contracts, they begin with a long list of definitions of terms. Patents don't require that terms be defined, and this creates havoc when a patent comes into the courtroom.

Leaving terms undefined is such an extreme and overwhelming problem that patent litigation in the United States now has a special proceeding called a Markman hearing, named after Herbert Markman, whose patent was the subject of a Supreme Court case that was decided in 1996. The purpose of the Markman hearing is to allow a judge to decide what the words in the patent mean.

Inventor Beware

The biggest hidden enemy in any patent application are words or phrases that have multiple or ambiguous meanings. Many contracts start with a list of terms and their definitions. For some reason this requirement is left out of patent applications, which can be a problem during an infringement litigation. This can cost millions of dollars to deal with after the fact, and can lead to invalidation of your patent.

Read your application over and over before it is filed, and make sure you provide an exact definition for every word or phrase that an opponent could dispute.

For the inventor, this is an astonishing and upsetting process. Any word used in the application can come under attack. Does a word have two possible meanings? Does a word that has always meant one thing to you actually mean something else to another person?

The attack can come against the most common and seemingly definite words in your field. It drops you into a bizarre wonderland where a perfectly clear sentence in your description gets turned 180 degrees into just the opposite of what you intended, or rendered ambiguous and impossible to interpret.

The Supreme Court's decision in the Markman case was that this process was so complex and messy that no jury could reasonably make decisions about it. The judge was put in the place of the jury and a special "trial of the words" was created—now called the Markman hearing.

It doesn't have to be this way. At the start of your description, make a list of every technical word in your application and write out what you mean by it. As long as you do this clearly, you are entitled to make a word mean whatever you clearly say it means—technically this is called being your own lexicographer. It is definitive and binding in court.

In addition, generate a concordance that lists every word in your patent application, and attach it is an appendix to your application. Read through it and consider checking definitions of many of the words. Watch for words with more than one definition. In the appendix, you can clarify the meaning of ordinary words and phrases such as "between" or "varies" or "separate from," as these sorts of terms can sink your patent at the Markman hearing stage.

Patent Vocab

PHOSITA is an acronym that is widely used in the patent legal world. It stands for person having ordinary skill in the art. This is a conceptual person who is knowledgeable in the field of your invention but completely uninventive.

On the day you filed your patent application, the PHOSITA should have had no way of figuring out how to do what your invention does. Your written description in the application is sufficiently informative if a PHOSITA could immediately practice your invention once he she has read your patent.

The PHOSITA

The audience for reading your patent description is supposed to be a person having ordinary skill in the art. The acronym for this is PHOSITA and is often pronounced FOS-it-ta. A PHOSITA is a common, well-trained, experienced person working in the field of the invention who is not inventive by nature.

When you choose to use technical language, the words, terms, and method steps should have usual meanings and be well understood by the PHOSITA. That is the concept at least. In the end, it is attorneys and judges trying to put themselves in the place of the PHOSITA who really matter. In a dispute over the meaning of a term, the opposing side will produce an expert in your field who will testify that he has no idea what your word means. Your side will produce an expert who will testify that every single person working in that field knows exactly what you mean whenever they see that word or term.

This dichotomy can be astonishing to an inventor. In one patent case, I had the chairman of the Department of Neuroradiology at a major university testify that he had no idea what a nerve looked like in an MRI, whereas our expert testified that every neuroradiologist was trained to know exactly what a nerve looked like in an MRI. So when the patent application said to first identify a nerve and then do measurements on it, the argument was whether a PHOSITA could identify a nerve.

That neuroradiology doctor who said he didn't know how to spot a nerve would never tell a patient or his colleagues or a referring physician or a student physician that he had no idea how to identify a nerve, but he had no trouble stating this in exchange for tens of thousand of dollars of witness fees paid by a major corporation. In this instance, the comment may have seemed to be a good idea when the attorneys and the witness were planning their testimony, but it ended up looking silly and just undermined the corporation's position.

What the PHOSITA does or doesn't know or what the PHOSITA will or will not understand is decided by the judge. Definitions and appendices can help reduce the likelihood of this problem cropping up.

Another clarification you can write into your application is to specify what kind of person a PHOSITA is for your invention. For instance, suppose your invention is software that converts a poem in classical iambic pentameter (a form used by Shakespeare and many ancient Greek poets) into a poem in dactylic hexameter (used by Homer) or into anapestic tetrameter (used by Lewis Carroll). We can expect a poet to know what these terms mean. However, your invention is probably a complex software algorithm that no poet could ever be expected to understand. The software might be simple to read for a computer programmer who works in the field of mathematical software writing. But the vast majority of expert computer programmers who will be asked to give testimony will honestly say they have no idea what a dactylic hexameter is.

The patent law says your description should make sense to someone who makes and uses your invention. In this case if you are an expert computer programmer and a poet, you have both skill sets and therefore would be able to make this invention. In order for the court to find that you have described your invention sufficiently, you should say that this will be understood by a team that includes both a classically trained professor of poetry and a mathematical programmer with experience in language software. Describe the PHOSITA or PHOSITAs in your application and the infringer will be forced to respect your position.

Getting the Patent Application Filed

Although inventors often file their own provisional applications (as described in Chapter 7 and appendix), most inventors have a patent attorney or patent agent file the actual patent application for them. The application gets assembled in a particular order, and has a cover form that identifies the type of application. The fee is larger for the patent application than it is for a provisional filing, but there are fee reductions for those with small entity status and even further reductions, under the AIA 2011 law, for those who qualify as micro entities.

AIA: Change of Rule

Overall, most commentators believe that the AIA was a disaster for the solo inventor it was meant to help. One small benefit the individual inventor enjoys is reduced fees at the outset. The AIA established the micro entity, which has even lower fees than the small entity.

Filing can be done electronically. Your patent attorney or agent will have an account that allows him to file. The individual inventor can file some items electronically or in paper format without any special account. You also have the option of establishing an online account with the USPTO so that you can file forms electronically and get some tracking information as the filing progresses.

Small Entity

To qualify for small entity status, the inventor must be an individual or a small business that meets certain size limits, usually fewer than five hundred employees. In addition, small entity status is available for nonprofit organizations, universities, and certain tax-exempt entities.

If you have an obligation to assign all or part of your invention to an entity that is not a small entity, then you lose the status—unless you are required to assign to an agency of the US federal government. You have to submit this information with the initial filing on an appropriate form for your fees to be covered throughout the patent exam process.

You should notify the patent office if your status changes, for instance if a huge multinational corporation agrees to take an exclusive license to your invention before the patent is even granted and continues the patent prosecution (pursuing patent grant) on your behalf.

Micro Entity Status

To qualify for micro entity status, the inventor must meet the requirements for small entity status as well as the following:

➤ The inventor hasn't filed more than four previous provisionals, formal patent applications, or foreign patent applications as an independent inventor (previous filings don't count if the inventor was employed by an entity that filed on your behalf)

➤ The inventor's income (or the income of the small company's owner/CEO) is no more than three times the IRS-determined median household income for the year

➤ The inventor doesn't have an agreement to assign the patent to an entity that doesn't qualify for micro status

Review for Deficiencies

Once all the parts are completed, the drawings done, the forms filled out, the oath signed and notarized, the correct fees identified, and payment arranged, the application is ready to file and can be transmitted to the patent office.

Among the first steps taken by the patent office is to review the filing for any deficiencies. If you leave something out, the patent office will let you know so you can correct the deficiencies without losing your original filing date. Once the deficiencies are corrected, the examination process can begin.

Using the Patent Cooperation Treaty Route

You have two ways to position yourself so that you can go ahead with various international patent filings in the future.

The most common in the United States is to file with the USPTO, and then consider the initial findings of its search report. If the USPTO discovers or identifies a damaging bit of prior art, then you may decide to cut your losses and abandon the effort, or to limit your expenses to the US filing because you believe the eventual rights that will survive the prior art won't be all that valuable.

Patent Vocab

The Patent Cooperation Treaty (PCT) was enacted in the 1970s so that a US inventor could file one application that covers both US and international patents for member countries.

However, if the initial evaluation looks good, you can progress to file a Patent Cooperation Treaty (PCT) application within twelve months of the date of your initial US filing. The PCT office will then do its initial search as well. Then you have eighteen months to progress to whatever national phase application you select. You might then look at the list of about 150 PCT countries and select, for instance, Canada, Australia, Japan, and Europe (designating England, France, Germany, and Italy within Europe).

Your second choice is to file directly with the PCT. Then after you see the PCT's initial search, you make your decision (eighteen months after the PCT filing date) as to where to file nationally and include the United States on your list. This is the most economical approach of the two, although it delays your US filing by eighteen months.

CHAPTER 10

Inventor and Attorney Communication

In This Chapter

➤ Organizing information for your attorney

➤ The components of a patent disclosure

➤ Adding detail through patent appendices

To succeed with an invention and a patent, you need to have great communication skills at a variety of points during the process. Among the most important need for communication skills is when exchanging ideas with a patent attorney.

Communicating with a Patent Attorney

To explain your invention and produce the best possible patent application, you have to be able to consider the audience you are speaking to. Patents have three principal audiences:

1. The PHOSITA: Your scientific and technical peers must be able to understand the language and information in your patent.

2. Business people: Your patent must be clearly understood by business people so they can grasp what your invention may accomplish in a given market, and how it will provide a commercial and financially valuable advantage.

3. The legal people: A lawyer or judge must be able to clearly understand your invention and your application.

The Legal Mindset and Technology

It helps a lot that nearly all patent agents and patent attorneys have at least an undergraduate degree in an area of science or engineering. However, if the patent is ever litigated, the attorneys who do the patent litigation are not usually patent attorneys—in many cases, they have not passed the patent bar exam. They are litigators, a particular breed of courtroom warrior. Many patent litigators have technical backgrounds as well. Some even have doctorates in science, engineering, or medicine, but many do not.

Patent Vocab

Patent attorneys help write patent applications and prosecute them before the patent office (the process of trying to get the patent granted) and have a special credential beyond just going to law school and passing the state bar exam. They also have to pass the patent bar exam.

Those who take and pass the patent bar exam but are not attorneys are usually patent agents. They help draft and prosecute patents at a lower fee to the inventor.

A general attorney who has not passed the patent bar exam should not, and generally will not, agree to help you draft, file, and prosecute a patent application.

Most important to keep in mind is that judges normally do not have a technical background, so your patent application must be understandable to the layperson.

Your First Draft

Your task of producing a document that speaks to technical people, business people, and legal people starts with your presentation to your patent attorney, who will draft the patent application based on what you have to say.

A good place to start is to write an invention overview in the format described later in this chapter, and to write your own version of a patent application before your initial meeting with an attorney.

For a first-time inventor, this is an exciting moment. It is a transition from being a technical person into being an inventor who is functioning in a legal and business setting.

In this initial meeting and any follow-ups, you have the task of getting the attorney to understand your invention nearly as well as you understand it. If you can accomplish that, then the attorney can start helping you add to your description in ways that will strengthen the patent application document.

Anticipating the Day in Court

As you prepare your patent application, you need to remember to make it as fail-safe as possible in anticipation of the possibility of litigation.

If the patent is litigated, your opponent's attorney will try to convince the judge that no reasonable person reading the patent would be able to understand it. He will point out confusing phrases, incomplete explanations, words that have conflicting meanings. He will insist that the description has major gaps and contradictions. If anything your attorney writes is unclear, your opponent's attorney will tell the judge that the inventor obviously did not understand how to build a working device since the patent is just a cover-up for guessing.

What can you do about this? You have to struggle to put yourself in the shoes of your opponent. Once you have a working device or process, you should be able to describe how it works. That may sound simple, but consider what the opposing counsel may have to say about what was written.

In addition to the invention overview, you should start making a list of necessary technical words and their definitions. You need to take care to recognize that a word may have a clear precise meaning in your own mind, but that others may see alternate meanings in the same word.

Patents Expert Pointer

Once you have your list of technical words, it's a good idea to get two or three technical dictionaries such as the *Penguin Dictionary of Physics* or the *IBM Dictionary of Computing*, specific to your field or related fields, that have been published as close as possible to the date of your patent filing—this is going to be a "definition time machine." First, make sure that your use of words matches what is in the dictionary and then safely store the dictionary for that day 17 years from now, when you have to produce it in court to prove how a particular word was understood at the time the patent was written.

Refer to well-respected dictionaries, and for every critical word in your description, make sure that a dictionary defines the word or phrase in the same way that you do. You can attach this list as an appendix to your patent application or store it in a safe place. At any time in the future, you will be able to use the list to prove to the court that the words in question did have the meanings you said they did at that time you chose those words.

You can include definitions that match your views in the patent application, but in some cases they can be overturned if the words you use in the definitions are also subjected to attack.

You may be thinking: "OK. This is ridiculous. Here I have this fantastic invention, how can it be invalidated by a battle over dictionary definitions? How can a judge decide that an important standard term in our field is ambiguous?" Well, you had better get used to it—that's how patent litigation plays out.

If your search through a dictionary or textbook does not confirm your idea of a meaning, or if you find that what you thought was a standard term isn't mentioned in the dictionary, then you have just made a very important discovery that can save you a great deal of trouble later on.

Consider the term *a quick-releasing capture device*. Here, you meant to describe a grabbing instrument that has a ratchet clamp. You push a lever, the ratchet releases, and the clamp opens up to let go of what it's holding. The court rules that a single thing can't be doing two opposite functions—it can't capture and release at the same instant. Maddening yes, but it is your job to spot these kinds of issues while the patent application is being written.

Watch out for some typical missteps. You could provide troublesome words and phrases to your attorney in the course of describing your invention. Your attorney could take a clear phrase from your description and accidentally make it ambiguous by substituting a word or two. Always stay sharp and review both your version of the application and your attorney's to catch any unintentional ambiguities.

How Much Do You Need to Know about the Law?

The patent attorney or agent is your representative, and it is his job to know the legal details. However, it is helpful to do your best to educate yourself about the legal side of patents.

On one level, the simple fact that your time with the attorney will be limited—partly by expense—means that the time required for the attorney to explain the basics to you eats into the time that is available for exchanging technical information about your invention. Some of the most important legal concepts for you to learn about are reviewed in Chapters 11 and 12.

The legal issues fall into three major categories:

1. The rules that the patent office uses to determine whether you should be granted a patent

2. The way in which the language of the claims of a patent are written.

3. The kinds of attacks infringers can employ to try to destroy your patent if you challenge them.

Elements of an Invention Overview

You can get initial exposure to preparing a patent application by preparing an invention overview, or disclosure. This is an informal document that you write to describe your invention for various audiences. You can update and amend it along the way.

The patent itself—once granted—may be so full of technical and legal jargon that you may barely be able to understand it yourself. For this reason, when you want to provide a written explanation of your invention to business partners, investors, marketers, and colleagues, it is helpful to have a more understandable version of the patent that specifically addresses the various interests that others may have in your invention. This overview is also a helpful document to share with the attorney or agent who will draft your patent application.

There are no specific rules, but the following is a suggestion for the major parts of the description. It will probably end up being four to ten pages long, but this will depend on how complex the information is. The appendix has a sample invention overview.

Field of Use

The first part of the overview describes the general field of specialization in which the invention will be important. In the example in the appendix, the invention is in the field of medicine, specifically in the field of medical imaging. The document explains that within medical imaging, the invention relates to the use of imaging and other diagnostic methods for nerves.

Purpose

Next, you begin to explain what your invention does. This is the utility or use or function that it accomplishes. In part, you have to explain what is missing in the field to be able to explain why the invention is beneficial. It might accomplish something entirely new; it might accomplish something that is already widely available but in a faster or more efficient way.

Existing Methods

In this section you discuss similar devices and methods that come close to your invention, pointing out their boundaries and limitations. This is your opportunity to mention the devices and methods that will be competing with your invention. You will be describing the state of the art or possibly the prior art in existence at the time of your invention. You need to research this carefully. You may believe that nothing comes close to your invention, but are you sure? Do you believe that nothing matches your invention no matter how far and wide you search?

To cover all your bases, you can mention completely different devices that are used to achieve a purpose similar to what your invention achieves. The example in the appendix of an MRI method for examining nerves by medical imaging could include mention of electrodiagnostic methods that send electric shocks into nerves to determine if they are connected properly. This allows the inventor to suggest that if an MRI scanner could see the nerve, then the existing electric shock method could be avoided.

This argument is important when you point out that your method is nonobvious. The basic idea is since we need to be able to test that nerves are properly connected (not damaged or cut), and since the best we can do is send in crude, painful electric shocks, and since nothing more effective and less invasive has yet been invented, your invention is nonobvious. You close the deal by pointing out that there has been a desperate need for your invention, understood by many, and the one reason it was never solved is because no one else had had the necessary flash of creativity needed to make this nonobvious invention. If it had been obvious, then dozens of PHOSITAs would have invented this improved method long ago.

Novel Features

This is where you carefully point out the ways in which your invention is new and different from everything that went before it. Some inventions have a single simple advantage or difference; other inventions may incorporate several advances. You need novelty to get a patent. You can also explain how the novel features result in new and improved results compared to what went before.

How it Works

Now you have to fully explain how your invention works. You can see in the medical imaging example in the appendix that the documents have to explain how a standard MRI scanner works and identify all of its parts before the description can progress. Only once this groundwork is laid out is it possible to go on to point out how the specific new, or novel, aspects of using an MRI scanner for the advanced invention in nerve imaging will work.

It is more difficult to explain why patent law requires inventors to explain something that every PHOSITA in your field already knows. It is best to think of this background as a necessary narrative for the attorneys and the judge. It should take them from what any generally educated person would know, and lead them up to the point where you are ready to specify the detailed way in which your advance fits in.

Some inventions require more basic explanations than others. Your patent writer will have to decide correctly how much to include.

Inventor Beware

There is no real penalty for providing too much detail in the text and appendix of your patent, but there is a huge penalty for providing too little information. By writing your own appendices, you can avoid the self-destructive syndrome of leaving out critical information in order to limit the expense of your patent attorney's time.

The problem with leaving out information is that the opponent's counsel in any future patent litigation may insist that your invention is not fully enabled because you have left out parts that no one could guess at with certainty. By specifying a complete example, you fully enable the invention. The patent may then cover a range of variations of different types of MRI scanners (for instance) that you have not described, but what is needed is a description of at least one complete system, including your novel part, that actually does work.

Known Uses of Similar Techniques

This is the section that focuses on the details of your innovative steps—the changes you have made to the technology. Here you want to show that the kind of changes you have made may have been used in very different settings or with different uses, but that nothing in those uses relates too closely to what you have done.

To understand this, it is helpful to consider the advanced technology plow system used as an example earlier in this book. It uses GPS to do the guidance and seed spacing but is innovative in that it uses a combined laser and ultrasound system to identify rocks in the soil ahead. The data from the laser and ultrasound systems is used to guide a robotic arm to remove the rocks or to lift plow sections over the rocks. There is a lot of technology there, but the invention focuses on using the laser and ultrasound system for rock detection.

In this part of the overview, you might point out that laser and ultrasound systems have been used before to verify the position of brain structures during neurosurgery and to guide welding equipment in an automobile assembly plant. However, it has never occurred to anyone that if the comparison of the two data streams was done correctly, it could locate rocks ahead of an advancing robotic plow.

You could also point out here that in the brain the systems find the fluid ventricles (low density areas), and in welding they're used to locate the edges between two metal sheets. In neither of those applications did it occur to someone that it could be used to identify hard solids within flowing solids.

Status of Research

It is helpful for the patent application drafter to know exactly where things stand in the development process of your invention. Early on, you might have said that the concept has been worked out and that the process is almost working and should be operational in a few weeks. The patent attorney may suggest holding off on the filing until it is actually working.

The danger of filing too soon is the possibility that the description will leave out a critical step that hasn't been discovered yet. Without that step, the process might not work. Any future opponent can point out that your patent describes something that doesn't work and therefore the patent is worthless.

If you wait to file till at least one version is fully operational, then this section might indicate that you now have five scientists testing various alternate uses and applications of the invention. In addition, you might indicate here that several similar new methods are now operational as well. You could show some of the actual results or products in some attached images.

Status of Filings and Disclosures

This section is a good place to describe what has been done so far to protect your invention. You can list and identify any provisional applications you have filed and any lab notebooks or signed witnesses' disclosures you may have prepared. Any foreign filings have to be mentioned here as well.

Disclosures to anyone outside of a confidentiality agreement are a huge problem. There should not be any. This is a change from 220 years of US patent practice. The AIA converts the US patent system to first inventor to file (see Chapter 6). This means that most of the time, it's irrelevant that you can prove you thought of something before someone else did. The patent office is only concerned with who filed for an invention first, not who was the first to think of it.

There is still a place for the sworn witness lab notebooks. They do help prove that you really did come up with this invention by documenting the steps you went through to arrive at the invention. Also, the AIA 2011 Act does include a means of protecting yourself against derivation, which this is when someone gets an idea from you without your permission, and then files for a patent before you do.

You need to prove two things to stop someone from stealing your invention:

1. That you invented first

2. That the other person actually got the idea from you

The documentation can show the date on which you had your idea, but may not be helpful for proving that someone else copied it.

Patents Expert Pointer

Even though you can no longer win a conflict with a competing inventor by relying on a signed and witnessed notebook to prove the date of your invention, it may still be useful to follow this practice. You will be able to prove the time and sequence of events you went through to arrive at your invention. This could be helpful if someone later tries to claim that you merely stole the invention from him in a derivation attack on your legitimacy as the inventor.

In older US patent law, you could publish a report or give a talk at a meeting or other public presentation and still file for a patent as long as you filed within twelve months, called the grace period. The moment you made a public presentation, however, the invention became unpatentable in nearly every country in the world except in the United States. Europe is out, Canada is out, Australia is out, Japan is out—none of these countries offers a grace period, and there is nothing you can do to fix the problem once the error is made.

Going public before filing usually happens before an inventor realizes the idea is patentable or under the mistaken belief that certain disclosures are safe or OK.

Of course, each page of the invention overview should have the word Confidential on it, and you should keep the document away from public view. Overall, once you have filed the patent application, then you can publish and discuss, but not before you file.

The grace period in the United States is now nearly useless because if someone hears your presentation of your new idea and files first, you have to prove that he heard your presentation and hadn't thought of it before they heard it from you. If you can't definitely prove he heard your talk or read your report and got the idea from you, then he wins and you lose, even though you invented first. It is better to act as if there is no grace period and never disclose until you file.

In this section, therefore, you lay out everything you know about documentation, disclosures, confidential discussions, and any preliminary filings. The patent attorney or agent will be able to tell you how to make the best of the situation if you have made a nonconfidential disclosure and whether it is salvageable.

Ownership

You need to lay out everything you know about any employer, university, granting agency, government, or any other entity or person who might have a claim to owning your invention. Keep in mind that many employers require employees to sign their inventions over to them (see Chapter 5). You may still be able to file for a patent when someone else owns it. It is also the case that the owner may be able to file with little participation by the inventor.

Business Prospects

In this section you lay out your thoughts on the business potential of your invention:

➤ What the price would be—based on the structure of the market and the pattern of other products

➤ How many units would likely be sold

➤ What the market looks like in terms of competing products

This will help you decide if it is really worth going through all the trouble of patenting.

If it does seem objectively possible that your invention would be of significant value, you can make secure, confidential contacts with various potential investors. If you have filed a provisional application, you may be able to get them to look at your invention overview so they can gain an understanding of what the invention is about. Note that in some cases, you can get investors to sign a confidentiality agreement to look at the overview, but that many companies will not sign one and will just tell you to come back again later after the patent application is filed.

CHAPTER 11

Understanding the Language of Patent Claims

> ### In This Chapter
>
> ➤ The different types of patent claims
> ➤ Why patent claims are so complex
> ➤ How language strengthens a patent application

Compared to the main text of the patent application description, the claims are complex and seem to be written in a special language for patent attorneys. After all, most are written by patent attorneys or agents. You may read some claims in your own patent over and over again, trying to decipher exactly what they mean and how they relate to your invention. But most inventors don't pay a great deal of attention to the claims of their patents.

There is a legal fiction, though, that the inventor has written the claims and the PHOSITA will be able to understand them. As difficult as it may be to understand the legalese, it's important for you, the inventor, to make sure the claims are correct. If the patent attorney went off course and left out something important or added something that goes beyond what was actually invented, then there is at least a possibility of fixing the errors if you are able to read and evaluate the claims.

What is a Patent Claim?

Although there already is a detailed description of your invention (see Chapter 9), the claims are specific legal statements of what makes up your invention. The claims cannot go beyond what you describe in the text of the patent specification. However, they describe specific parts of the invention.

You can have just one claim, or there could be hundreds of claims, depending on the nature of the invention. Each claim describes a slightly different aspect of the invention.

Stating a Claim in Just One Sentence

One reason that claims can be so difficult to read is that there is a formal requirement that each claim be a single sentence. The result often is an extremely long and grammatically complex sentence, including lists and various levels of sub-sentences. The effect of the single sentence rule has been to further formalize the way the claims are written.

Patents Expert Pointer

Early patents did not include claims. The use of claims was introduced in the US patent system in 1836, but they did not appear in the European system until the 1970s.

Claims used to be simple and just stated that the inventor claimed what was described in the description. Now, however, there is a much more important role for the claims. Even if the description is detailed and effective, you can find that your patent is useless if your opponent can prove that any claim is indefinite, which more or less means ambiguous.

The imperfect idea here is that a claim on a piece of intellectual property should be more or less like a claim on a piece of real property on the ground. The claims should provide the "metes and bounds" of what is claimed, just as the surveyors' measurements mark out exactly the boundaries of a piece of land.

Specificity of a Claim Sentence

Claim sentences generally start with a preamble that states some general matters about the field of use and existing known components. There is next a linking phrase, and then the main body of the claim, which uses language to generate a line in the sand, but in terms of the invention.

For example, you might have said in your description that you have invented compound *C*, which results when compound *A* is mixed with compound *B* and then heated to a high temperature and then cooled abruptly. In the text, you might further specify in your example that three parts of compound *A* are mixed with two parts of compound *B* and heated to 200 degrees Fahrenheit, then cooled in five minutes. That is the best mode, and it is fully enabled—it works.

Does this mean that a potential infringer can get around your patent by doing exactly the same thing, but heating to 201 degrees? In your claim you might say that it is heated to at least 180 degrees but no more than 220 degrees. You might have discovered that at over 210 degrees, compound *A* undergoes a transformation so that when the mixture is cooled, you don't get the desired compound *C* but rather just a degraded mess. By claiming boundaries, you prevent someone from simply stepping around your patent.

If you just say it is heated but don't specify any actual temperature, the court might find that your claim is indefinite and discard the claim. Here the argument is that the PHOSITA would have to engage in "undue experimentation" to learn how to make *C* according to your method.

But the attacks from the opponent do not necessarily stop here. They can further go on to say the claim is indefinite because it wasn't stated what sort of thermometer should be used and whether or not someone would be infringing on the patent if he heats it to 179.8 degrees.

To avoid this, the claim can specify how a measurement is made, or it can refer to a specific method of measuring temperature provided in the description. It can also specify the usual degree of accuracy of the temperature measure to be used: "measured to tenths of a degree and rounded according to the rules provided in the *Penguin Dictionary of Mathematics*

Inventor Beware

Remember that in patent drafting, ambiguity is one of your biggest enemies. Every critical phrase—even if common words are used—has to be carefully repeated and thought through. You want to be the one to discover an ambiguity before your patent application is filed. The alternative is that you may learn about the ambiguity when an opponent points it out to the judge as grounds for invalidating your entire patent.

under algebraic rounding."

You will need various claims and claim components to describe how much of compound *A* will be claimed relative to how much of compound B. What temperature will it be cooled to and how fast will the cooling take place. Are there other similar compounds D and E that form F in a similar way?

Independent and Dependent Claims

As a means of shortening and simplifying claims, the patent rules allow for writing independent and dependent claims.

An independent claim has all the details laid out, such as the temperature example about compounds, *A*, *B*, and *C* in the previous section. Let's say that this information is common to all the different variations you can imagine—mixing this approximate ratio of ingredients, and then heating to between 180 and 220 degrees to the nearest tenth with a mercury thermometer, rounding according to standard rules. Let's say that this information appears as claim #1.

For a dependent claim you could now write, "The invention of claim #1 wherein the cooling is done to between 20 and 40 degrees Fahrenheit, using a mercury thermometer to tenths and rounding." This would be your claim #2. You can then write, "The invention of claim #2 wherein the cooling is done in less than five minutes and more than two minutes," and call that your claim #3.

You can now limit your original heating claim a bit with a further dependent claim #4: "The invention of claim #2 wherein cooling is only done in less than four minutes and more than three minutes."

What has happened here? Claim #4 has several of the steps lined up, but it only covers a smaller range of cooling times. This is a further limitation. One reason this is done is that someone else may have previously made a similar compound but specified five minutes of cooling. You are now claiming a different invention that is outside the bounds of what the other inventor claimed.

Types of Claims

It is often the case that a given invention can be claimed in several different ways with regard to type of activity or function in the claim.

Suppose you have invented a machine that can draw short line segments in various colors and lengths as a response to music. The length of the lines reflects the tempo and the length of the notes; the color reflects the frequency or tone. This results in a unique and attractive drawing that is recognizable and particular to each song.

You can use a product claim to describe the apparatus or device that does this. You can also claim the process of converting songs into colored line segments (independent of the type of device that does it). You could also claim a colored drawing type that is made by applying colored line segments the length and color of which are determined by music analysis.

Patents Expert Pointer

The patent expert thinks of several different ways to claim the same invention. Think of your invention as a device that can be made, or a process that can be followed. Is your invention a product that is unique because of the way it is made? The more versions or views of your patent that you can imagine, the better.

Means plus Function Claims

One particular type of claim that was designed to help make claim writing more efficient provides an example of what can happen in patent law over time.

The idea of a means plus function claim was to use a small number of words to claim a large number of devices or device variations. Using our above example of a method of representing music as colored lines, the means plus function claim would claim an apparatus as follows: "A means for converting music into colored lines wherein the length of the lines reflects the tempo and note length and the color is controlled by the frequency." Notice that the claim doesn't really say much about the device. It just says "a means for" although it does describe the function that the device will accomplish.

In a standard device claim, the wording might be: "Microphone with an output into an analog to digital converter that sends its output to a signal-processing chip that performs Fourier transforms on the digitized waveforms, having all the instructions in ROM hardware and listed in the appendix of this patent, wherein the numerical representations resulting from the signal-processing steps are inputted into an 8086 or similar microprocessor with ROM memory that uses an algorithm in an appendix to this patent to assign numerical values from the music data into line lengths and color selections that operate a standard digital plotter device, for converting music …."

In this version, you can see that the invention is described with a great deal of particularity. The information should be sufficient for a PHOSITA to be able to build one of these.

However, this claim is fairly specific, and it might be easy for someone to design around it. Perhaps the signal-processing step and Fourier transform could be skipped.

There is a law of equivalents that can get you protection for similar or equivalent devices based on the fully written claim. However, the means plus function claim was supposed to get all this detail out of the claim section so everything could be explained in greater detail in the main text of the specification.

The means had to be spelled out in great detail in the description. This also provided for the fullness of language, drawings, block diagrams, and extensive appendices to build the case for a wide variety of well-described means for accomplishing the claimed function. This was intended to reduce the difficulty of packing long, complex descriptions into the single-sentence claims, while also giving space to allow for the drafter (writer) of the application to explain the meaning of the various means.

The problem perceived by those being sued was that the means plus function claim was an overly broad claim, sweeping up a wide array of devices and methods, and making them infringing devices or methods without explaining them.

In a series of court decisions, the patent law was progressed so that if a means plus function claim was used, then the requirements for spelling out all the details were far more demanding than if the description appeared in the claim in its full form.

One example of the tightening of requirements is the standard technique of referring to an existing academic paper or patent application that explains a particular prior art algorithm or machine in great detail.

Traditionally, it was possible to make a statement in the description such as, "we used the Johnson algorithm for this step as detailed in the *Journal of Computing Science*, volume 153, page 25–34, "Details of the Johnson Algorithm," by Tom Edwards (1995). However, with the updated rules on means plus function, it was no longer allowed to reference an outside publication and therefore rely on its contents. It was as if the PHOSITA was now disabled from referring to the literature. Although this does not necessarily make a great deal of sense from a theoretical point of view, that's the law.

So if a means plus function claim is used (even if it was written in 1994, long before the new rules came into force), then any information from an external reference had to appear within the "four corners" of the patent. This means it had to be written out in the text of the patent description or included in an appendix.

As the rules became hopelessly difficult to meet, this changed from a type of claim into a "defense against infringement." Infringers would try to convince the court that a claim was actually a means plus function claim even though it didn't use those magic words. If the court accepted the argument, then the description and outside references in the text would no longer meet the standard, and the patent could be invalidated for indefiniteness or lack of clarity in the claims. The use of means plus function claims and related step plus function claims (relating to software and methods) have mostly been abandoned by writers of patents.

Guidelines for Claim Writing

As with every other aspect of writing a patent application, there are rules that need to be followed when writing claims.

Definitions Are Paramount

To keep claims compact and make their meanings precise, it is important to analyze every word, every phrase, and every section of the claim to identify possible areas of ambiguity.

There are two kinds of problems with words:

1. Technical definitions: In the patent application, you can be your own lexicographer and clearly define specific technical terms you will use. It is actually advisable to define the same word twice using different language because tremendous attention can be focused on a definition in an infringement case. Any ambiguity in a definition can be fatal to the patent. Often the risk in the definition arises from the second kind of definition problem.

Patent Vocab

The word *lexicographer* appears in the phrase that "an inventor can be his own lexicographer." It means that you can provide any definition you want for a technical term in your patent. As long as you take the time to write out very clearly how you want a particular word to be understood, then your definition trumps any standard usage in your field, any dictionary, or any legal argument.

2. Common ordinary words: A famous example is a patent battle over a tennis racket invention where the claim used the term *varies between* and the judge decided that it could not be determined if this meant that the position alternated between two fixed positions or whether it referred to a variety of positions (at least more than two) that were between two extremes. When something passes through an area, is it in that area or has it gone past that area?

These are maddening ambiguities that can be in front of your eyes without being seen until your opponent wants to invalidate your life's work because we can't agree on what the word *between* or *through* means.

The problem is that language is not as precise as a mathematical formula. Words, phrases, and sentences carry meaning in complex ways. However, the claims in patent applications must be precise. If a claim is not clear to you after ten readings, the problem may be with the claim. Don't assume that the problem is just your lack of familiarity with the legal complexity of claim language.

Belt and Suspenders Wins the Day

One very helpful method of guaranteeing that your meaning is clear is to describe the same aspect of an invention in two or three different ways. Each description should be equally clear and precise, yet you should stretch your imagination to consider several entirely separate ways of saying the same thing. This is done because it is virtually impossible to predict how your description will hold up against an opponent who is represented by one of the world's leading litigation attorneys who is pleading before a judge who is suspicious of patents in general and doubtful about your patent specifically.

Claiming the Same Invention in Different Ways

In addition to using several different descriptions of the same critical feature within the claims themselves, it is helpful to claim different aspects of the same invention. Here you will need different sets of claims. This corresponds with the point made above about claiming a method, claiming an apparatus or device, and claiming a product by process. These can each be different aspects of your invention. Claiming different aspects of the same invention can be helpful because an infringer may try to circumvent your patent by duplicating an aspect or manifestation that was omitted in the claims.

The Benefit of Independent Claims

One reason that most patents have a series of independent claims is that this approach offers a defense against a future discovery of similar prior art that the inventor is unaware of.

This is the situation where you, the inventor, believe you have identified every possible prior art reference and have carefully written your patent and claims to totally avoid any problems with the prior art. During a future litigation, however, the other side discovers an obscure reference that comes very close to an aspect of your invention.

If you had just one claim, the obscure reference could knock out a critical phrase, thus rendering the entire claim invalid. If that one claim is all that you have, then the entire patent is gone forever. But if the invention is parceled out into a series of dependent claims, as described above, then the effect of the close reference could be to knock out one or two of the dependent claims without destroying the whole patent.

This is also why there should be more than one independent core claim, each with a set of dependent claims. Each of the independent claims should cover a different aspect of the invention, and each should use a different set of words to describe it.

The Inventor's Role in the Process

It may seem obvious to you that the claims in a patent application are so complex as written that no reasonable person could expect the inventor to fully understand them. These are written by experienced attorneys who have passed the patent bar exam, and you are an engineer or plumber or painter.

Patents Expert Pointer

The inventor should become totally familiar with the meaning of each part of each claim of his patent application. It is tempting to think that the claim language is all technical information between patent attorneys; however, you may someday be held responsible for an exact understanding of the claims. So even though they are difficult to understand, you need to take time to fully appreciate what they say.

Nonetheless, a favorite trick of opposing counsel is to put the inventor on the witness stand and ask him questions that probe his understanding of his own patent's claims. It may seem unfair to you, but the jury will think this is entirely reasonable. If the inventor can't understand and explain his own patent application, how could the inventor expect a competitor to realize he was infringing (or so the argument goes)?

On the one hand, as the inventor you will have plenty of time to prepare for this moment in the unlikely event that your patent ever has to be involved in patent infringement litigation. (Only a tiny fraction of patents are ever used in litigation.) On the other hand, if that moment does come, and if you are fully prepared to be comfortable with the language used in that future court room, will you be pleased with the outcome?

Years down the line, it will be too late for you to realize that the patent attorney didn't quite get it right. As the patent is being written is the time for you to have a rock-solid understanding of each aspect of your invention that is being claimed. You need to fully understand each independent claim and each dependent claim so that you can do your best to assure that it is totally consistent with the invention as you understand it.

A. G. DAVIS.
ELECTRIC LIGHTING SYSTEM.
APPLICATION FILED MAR. 4, 1899.

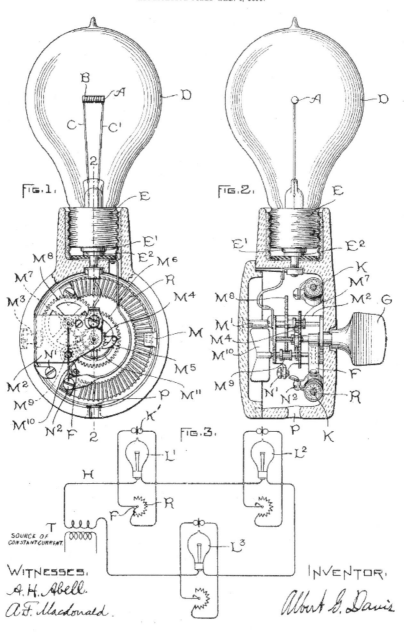

WITNESSES.
A. H. Abell.
A. F. Macdonald.

INVENTOR.
Albert G. Davis

CHAPTER 12

The Steps in Getting a Patent Granted

In This Chapter

➤ Determining whether an invention is patentable

➤ Notices and actions

➤ Turning a loss into a win

The day finally arrived when all the i's are dotted and the all the t's are crossed—you have submitted your patent application to the patent office. What happens next?

History of the Examination Process

The first person to develop the answer to that question was Thomas Jefferson in 1790. Although he was a prolific inventor, he was skeptical about patents. Nonetheless, as Secretary of State, he was assigned to deal with the very first US patent applications—the history is described in Chapter 2.

Jefferson's doubts and concerns led him to develop an examination system to evaluate new patent applications. This is not how it has been done everywhere.

The alternatives around the world at various times have included non-examination systems. In that arrangement, the applications are checked for completeness, filed, indexed, published, and granted. In a non-examination system, the quality of patents is determined only if a given patent gets used to try to stop an infringer. Everything is handled by litigation.

Jefferson was concerned that many worthless patents would be granted. Therefore, he personally read every one submitted for the first two years of the patent office's existence. The experience led him to develop an approach to examination of patents.

True, Congress dumped Jefferson's idea and freely granted patents without examination for a few decades after he passed the job along, but by the 1820s it had become apparent that Jefferson was right. So the patent office was established with an obligation to examine patents before granting them. Gradually, every other country has adopted an examination system, although many European countries did not start examining patents until the 1970s.

Ultimately, the main purposes of the examination process are to weed out patent applications that have readily identifiable fatal flaws, improve the quality of patents that are granted, and reduce the burden of litigation. The result of the examination is a prosecution history document that shows the various questions and objections raised by the USPTO and the responses that convinced the examiner to allow and grant the patent. By reading this prosecution history, a potential infringer can better understand the strengths and the detailed underpinnings of the invention from the point of view of patent law.

The Patent Examination Process

A patent goes through a lengthy and thorough examination process before it's granted. The process is broken down into steps.

Categorization

The first step in the examination process is the categorization of the patent by subject matter and field of specialty. This leads the patent office to assign your patent application to an individual examiner who will work with your application all the way through to grant or rejection. The whole process usually takes at least two years in the United States. In some countries, more than fifteen years can elapse between application submission and grant. It is not a quick process.

Completeness

The second step after categorizing the field of the patent and assigning it to an examiner is for the examiner to look through your filing to make sure you have met all the formal requirements for the filing. The components of a patent application are discussed in Chapter 9. These are what the examiner is looking for.

If any parts are missing, if the forms aren't completed properly, if the fee is not paid to the correct amount, the examination goes on hold and the examiner issues a deficiency notice.

Usually, it doesn't take a lot of time to correct a patent application. It keeps its original filing date even though it is not complete. If you respond to the examiner and fix the problems in the allotted time, then the examination process moves forward in earnest.

Prior Art

The first major hurdle your patent must pass is the patent office's search for prior art to make sure your invention is really new.

You can't get a patent if you are not the first true inventor of the invention. This dates back to the birth of the modern patent concept in the Statute of Monopolies of 1623 in England (see Chapter 2). We don't want to reward someone for copying someone else's idea. We don't want to give a monopoly to the second inventor and leave the first inventor with nothing—that doesn't seem just and it doesn't seem to provide an incentive for inventing new things.

So, to meet this requirement, the patent office conducts a prior art search along the same lines described in Chapter 8 of this book. The result of this search is usually a report in which the examiner lists five or ten publications such as academic papers or patents that come close to how you describe your invention in your application.

In addition, the inventor is required to turn over the results of his own prior art search to the patent office. If it is later proven that the inventor knew of some relevant prior art but did not provide this to the patent office, it is considered fraud and will allow an opponent to permanently render your patent unenforceable. There is no requirement that the inventor do an excellent search, but anything you know of that you do not disclose to the examiner is a fraud on the patent office and also violates the oath that the inventor provides as part of the application.

The examiner writes a letter designation next to each cited reference in the search report. There are a number of different letters used that signify the status or importance of each reference from the examiner's point of view. The most common and important designations are usually A and X.

An A reference simply provides background and context and helps define the state of the art in general at the time the invention was made. An X reference is trouble. This means that the examiner believes that the contents of the reference is so close to your invention that your idea cannot be considered novel or cannot be considered to show an inventive step.

You will have to respond to the search report by explaining to the examiner why he is wrong about any X references that turn up. This is routine, and it is usually possible to overcome these references by showing subtle or fundamental differences between your invention and what appears in the reference.

The prior art search can prove that no one has published your invention before. But the examiner also has to be satisfied that your invention has not been used or sold previously.

Priority refers to an even closer question. What if someone else invented the same thing and has kept it completely secret but arrived at the patent office to file his invention the day before you filed yours?

In the past, it didn't matter if another inventor filed first, as long as you were able to prove that you had conceived of the invention before the other contender conceived of it. Under the AIA of 2011, however, what matters is who filed first. It no longer matters if you thought of the invention first or developed it first. There is only one type of evidence that can save your invention if the other inventor filed first: you have to prove that the other inventor derived his idea directly from you.

AIA: Change of Rule

As long as a contender invented independently, it no longer matters who invented first; it only matters who filed first. This is even true if the other inventor files on the same day as you do, but has a time stamp on the application that is a minute earlier than your time stamp.

Nonobviousness

The idea of nonobviousness relates to a special concern that Thomas Jefferson brought up in 1790: would patents be used to protect trivial advances? The purpose of patents was to encourage inventive leaps that applied human creativity and imagination to advance the technological arts in surprising and unpredicted ways. This concept was meant to stand against new products or methods that represented small or obvious changes. These smaller or obvious advances were considered to be more or less inevitable.

This is still a controversial issue. Once someone has invented something, humans tend to immediately incorporate the idea into their thinking. It then seems obvious to everyone. This is a problem of hindsight. Almost everything new seems obvious after you understand it.

Determining Nonobviousness

The modern tests used by the patent office to determine nonobviousness were defined in a 1966 Supreme Court case dealing with an improvement on a plow from an inventor named William Graham who sued the John Deere company for patent infringement.

The original John Deere was an inventor in the early 1800s who made a major advance in the design of plows. The patent that the Supreme Court dealt with concerned Graham's invention of a new way to mount the blades of a plow. Graham was suing over his second patent, which was a modification of the spring mount from his own previous patent. The Supreme Court decided to invalidate the second patent, and in its decision it provided specific guidelines to determine if an invention was nonobvious.

Prior to the Deere case, the determination of "inventive step" was based on a general assessment that nothing in the prior art would suggest to the PHOSITA that he should pursue this invention and that nothing would have clearly predicted to the PHOSITA that a new invention in this area would "have a reasonable chance of success." The evaluation was supposed to look at the facts and information available at the time the invention was conceived of.

The Inventive Step

The general idea is explained by the following situation:

Let's say that ten companies are making gas pedals for automobiles and everyone is complaining that the pedals are so short that they are painful and push into the foot as the driver operates a car. Over the years, the various manufacturers have gradually progressed from a pedal that was a round 1-inch disk, to a broader pedal, and then to a longer pedal. Most pedals are now about 3 inches wide and 4 inches long. In the previous five years, pedals were just 3 inches long, but people are still complaining even with 4-inch-long pedals, although the width issue seems to be solved.

One of the manufacturers now comes out with a pedal that is 5 inches long and seeks a patent claiming that this required an inventive step. That patent is likely to fail because the patent examiner will probably say that the prior art suggested changes in the direction of lengthening the pedal. It had already been learned that widening the pedal eventually reached a width that everyone was happy with. It had already been shown that progressively lengthening the pedals improved their acceptance and reduced the number of complaints, but that many users were still dissatisfied. A theoretical noninventive pedal designer would follow the trend and lengthen the pedal further and would be able to predict that the longer pedal would function adequately and would probably be even more comfortable for drivers.

Secondary Considerations of Nonobviousness

Two inventions that help point out that even something very simple can be nonobvious are the window envelope and the Post-it note. These two inventions are convenient to keep in mind as we consider the three secondary considerations or indices of nonobviousness that were established by the Supreme Court in the *Graham v Deere* case.

One secondary consideration we can look at is commercial success. If the new product is successful in the marketplace, then the new invention is a useful change. The argument flows from the fact that if the new product is valuable and if it was obvious, why weren't any manufacturers making it earlier?

This particular criterion is usable only if the new invention is already on sale and has had time to become successful. Therefore, it is more useful for protecting an invention during a patent litigation. However, some inventions enter the marketplace just after the patent application is filed, and so the inventor can rely on this consideration if the invention is an immediate commercial success.

Another measure is long-felt need. Here the inventor argues that the invention solves a problem that everyone knew about. The fact that the problem, or the hardship resulting from the lack of the invention, was well known means that there must be some reason no one had offered a solution to the problem. The inventor then makes the point that everyone knew there was a problem, but that there was no obvious solution. This makes the case that what was required was an inventive spark of genius to make the intellectual leap to accomplishing a solution to the problem.

Commercial acquiescence is the other consideration. It captures the concept of respect of competitors for the advance. This is more complex to apply because it asks whether competitors naturally recognize the invention as patentable and one they won't be able to invalidate in court.

The evidence for nonobviousness would be in the offers to license the invention and efforts to invent around the invention. The evidence for obviousness would be in the tendency for competitors to offer the new design themselves, making an assumption that this change can't really be a patentable invention. This assumes that decisions to infringe are based on a patent examiner's view of technology; however a businessperson's decision to infringe may result from very different considerations. This is also a measure that is difficult for a patent examiner to make use of because it reflects the situation after the product goes on the market.

There are three additional objective factors that were developed by the Federal Circuit Court, which considers appeals of patent cases:

1. Evidence that others in the field were working in a different direction, teaching away from the direction taken by the inventor. This allows the inventor to argue that his invention succeeded by doing the opposite of what everyone in the industry would have expected to be useful or at least by doing something different from what others were considering.

2. Skepticism from experts in the field, who questioned whether the invention could possibly work. Even if the invention worked, the experts might predict it would be useless or a mistake.

A great recent example of this was the launch of the Apple iPad. When it appeared, there was widespread derision in the technical press and the newspapers. The New York Times ran a story reporting that everyone was agreeing that the choice of the name iPad was a disaster and represented a rare misstep by Apple. They said that the name evoked a feminine hygiene device. The humor about the name was paired with numerous articles by experts saying that no one would want one because they were completely useless.

Of course, the iPad was the most explosively successful launch of any product of any type in history, and it totally transformed the computer technology industry—one more monument to the inventive genius of the late lamented Steve Jobs.

3. Opposite of the commercial acquiescence test, is copying. This is another test that makes more sense in the setting of patent infringement litigation. The idea is that the fact that a competitor risked a lawsuit in order to sell the copied invention proves that the invention must be important and difficult to invent around.

One other closely related area of assessment concerns what is called a combination invention. Here, one is putting together two things that are already known in order to make something new. If the two parts each produce exactly what they did when they were separate, it is considered to be likely to be obvious to put them together.

Synergism

To be nonobvious, something unique, useful, and new should result. This is called synergism. The inventor should be able to argue successfully that by looking at the two components separately, the PHOSITA would not have predicted the beneficial qualities of the combination.

Inventor Beware

You can't get a patent for simply putting two things together that have not been put together before. The result has to be greater than the sum of the parts. A unique, surprising, and useful product must emerge that no one would have predicted would result from the pairing or assembling of components.

The Post-it note is a great example of this. Glue was known. Paper and notes were known. A Post-it note is nothing but some glue on paper. Most notepads are bound together by an adhesive spine. However, the use of a reusable glue applied as a strip along one surface of the note by Art Fry in 1974 created a wonderful new product.

Here we have a synergism. Given the spectacular number of Post-it notes sold since its invention, it is clear that if this invention were obvious, then someone who was selling notepaper over the many preceding decades would have already developed Post-it notes. The invention seems obvious in hindsight, but we have objective evidence, which can be framed in the tests above, that it must have been nonobvious.

The window envelope combined an envelope and a hole in a piece of paper, two things that were obviously well known. However, the particular way these two were put together, having an opening over the address field so one only had to type the address once (on the document), saved millions of hours of work in retyping addresses on the outside of envelopes and in matching documents to the their own specific addressed envelope. Why didn't anyone do this sooner? An inventive step was required, and no one made that step until Americus Calahan had the idea patented in 1902.

At a different extreme, some inventions are so dramatically unique and apparently complex that they seem almost magical, and no one raises the question of nonobviousness.

The Diffusion Anisotropy Imaging invention, covered in the patent in the appendix, produced a stunning new view of the internal architecture of the brain (see the appendix). It was widely known in that field that it would be useful to be able to make a medical image of the internal architecture of the brain. It was also obvious as to what the result would look like: it would show the neural tracts running through the brain. It was even known that capturing images of the property called diffusion anisotropy could be used to show the internal structure of the brain.

However, although thousands of the world's leading scientists and engineers in this field were aware of the need, how to approach the problem, and what the end product would look like, none of them figured out how to accomplish the result. Further, when the problem was solved, the provisional application was filed and the solution shown publicly, the scientific community failed to believe that the problem had been solved. Even after the solution was published, thousands of scientists struggled for five more years without ever producing a successful solution while the patent application was examined, published, and granted.

This proves nonobviousness because it is easy to show that a thousand scientists working on the exact problem without the necessary spark of invention could not see how to solve it. The result required putting together information from four very different fields that had not been combined before: neuroscience, MRI, computer graphics ray tracing, and vector mathematics.

The imaging example shows other measures of nonobviousness: failure of uninventive scientists and researchers to solve a problem, and elapse of time without matching discovery by others.

The elapse of time issue is an important one to keep in mind. The opposite situation is one in which ten different scientists make the same step at around the same time. This simultaneity of advancement suggests that the state of progress in the field had made the advance obvious.

Also, there existed extensive literature with thousands of publications in each of those four fields of knowledge. The simple fact that all that knowledge existed does not in any way make the invention obvious. In fact, with a large amount of prior art material, it is possible to argue that there was no way a person could predict which exact publications from the many thousands in the various fields should be selected to be assembled to make the invention.

Finally, one of the leading scientists in the field of MRI made a public statement after the invention was made, but before he realized it had been made, stating that the capability of visualizing the internal structure of the brain in this way was science fiction, showing doubt that the problem could ever be solved.

Enablement

Another major focus for the patent examiner is an attempt to determine if the invention is described in sufficient detail that it will actually work. This is a challenging problem for the examiner.

Originally, the US patent system solved this problem by requiring the inventor to produce a working model of the invention. This is a great solution to the problem, but by the mid-nineteenth century, it was already apparent that this solution was not viable. The problems with it included the variety of types of inventions that were difficult to represent in model form and also the difficulty of storing and managing all the models.

Patent Vocab

Enablement is the degree to which your written description actually tells a reader everything he or she needs to know to make or operate your invention. If you file the patent before you get it working, your description will probably not fully enable the patent. If you have the invention working but leave out a key step in the description, then your patent application will probably fail on lack of enablement.

The examiner does not go into a chemical laboratory to test out the chemical reactions described, nor does he try to assemble a copy of the mechanical invention. The best the examiner can do is try to be sure that the application has a detailed description of how to practice the invention.

Often, the patent examiner or supervisor will actually have a great deal of technical experience in a given field. In that case, the examiner may have a solid basis for determining whether the level of detail in the description is sufficient. In general, however, the examiner has to accept the inventor's assertion that the information is sufficient.

Later on, if an opponent in litigation can prove that a person of ordinary skill could not follow your instructions and make a working result, then your patent will be invalidated. You cannot add anything to the patent description after it is filed and the deficiencies are corrected. The claims can be expanded or rewritten, but the contents of text of the description are final. The inventor must include everything necessary as of the date of the filing of the application.

This is also why it is not a good idea to file an application before the invention is actually working. If the invention is not yet working in the laboratory, then you may be missing some critical final step you have not discovered or appreciated yet. In that case, even with a complete and detailed description, the text will be inadequate to fully disclose how to perform the invention.

Stages in the Examination Process

There are two different major routes used by inventors filing patent applications in the United States. The inventor may choose to use the Patent Cooperation Treaty (PCT) route or file directly with the USPTO. When a PCT filing is made, the inventor is choosing to start with an international scale approach. When a USPTO route is chosen, the inventor is choosing to start with the US office and to delay a decision on international filings. The examination process for the two routes is fundamentally different.

Taking the Patent Cooperation Treaty Route

If the applicant files with the PCT, then the first step in the examination process is limited to a search. At the end of a period of eighteen months after filing, the PCT office publishes the patent application along with the result of the search, including the A, X, and other markings for each identified reference that indicates how close or damaging the examiner feels the prior art reference will be (see above).

In theory at least, the publication of the application starts the clock ticking for competitors who can then see what new invention is coming down the pipeline. There will be no more secrecy. Notice that if the inventor chooses to abandon the patent application at this point, he has already published a full public disclosure of how to perform the invention that is wide open and accessible for the inventor's competitors.

At this point, the applicant can designate one or several national phases. In each of these phases, once the fee is paid, a different national office, such as the European Patent Office or the Japanese Patent Office, will start its own formal examination process. It is in this second phase where each of the selected patent offices will begin to evaluate the application for subject matter, indefiniteness, utility, novelty, nonobviousness, and enablement.

Taking the Direct USPTO Route

If a direct route to the USPTO is chosen, then the USPTO starts immediately with its examination. In the United States, there is no separate search phase and examination phase. The two steps are taken together.

If you also intend to pursue international rights, then you additionally file the same application in the PCT system twelve months after the US filing. Then eighteen months after that (thirty months from your initial US filing date), you enter the application into the various national (such as the European Patent Office) phases that you select. When this route is chosen, you will probably have a great deal of information from the US exam to help you decide whether to take on the additional expense of the various international filings.

Another aspect of filing in the USPTO is that if you intend to file only in the United States, you are allowed to indicate that you don't want to have your application published. If you select this, then the entire contents remains secret until the patent is granted. More importantly, if the patent is not granted, it is never published. The fundamental bargain of the patent system is that you fully disclose your invention in exchange for a monopoly. However, if your patent application is rejected, you could fully disclose your invention and get nothing in return.

Filing Outside the United States

For patent systems outside the United States, there is no choice on this issue—all applications are published in eighteen months by the PCT.

One of the advantages of publication is that it does make competitors aware that they may face the existence of a patent. This could lead them to stay away from infringing. Another advantage is that it makes it possible for competitors with prior art to notify the patent office and warn the examiner. This may sound like a disadvantage, but it gives you a chance to deal with an unknown reference early on when it is still possible to more easily amend the claims. One further advantage is that if your patent is granted, you may be able to assert some retroactive rights against infringers back to the date of publication.

National Security

The one other reason a patent application might not be published is if it is determined that your patent includes information of high value to national security. This might be a patent to do with an advanced weapons system or nuclear military technology.

This type of application may never be published even if granted. Further, you may be barred from filing internationally and from exporting or disclosing your patent on a confidential basis, even with a close ally of the United States such as Great Britain. Because of this issue, every provisional application and every formal application in the United States is subject to review for the need for an export license before a patent can be filed as a PCT application or in another national patent office.

Office Actions

The examiner first checks for completeness and requests correction of any deficiencies. Following that, the examiner does the patent search and conducts the first round of examination. It is usually the case that various objections and rejections emerge in this initial office action.

In this first action, the examiner assures that the patent application covers patentable subject matter (see Chapter 3 of this book) and also raises concerns about ambiguous language that could render parts of the patent indefinite.

Another general issue is utility—an assessment of whether the invention actually credibly does something. The classic example of an invention that fails on utility is a perpetual motion machine—a device that keeps working with no energy input, and which the examiner knows is impossible and cannot work.

Utility is not the same as usefulness or commercial value. An invention of a complex hat that has an umbrella, lights, solar panels, and a beer cooler with a straw leading down for drinking by the hat wearer may never sell a single product, but it does have utility—it does something.

The patent examiner needs to make an inference about whether the enablement is sufficient. This is a question of whether the description you provide is sufficient and complete so that an ordinary person in your field could read your application and then know how to perform or make your invention.

Aside from all of these preliminary issues, the core of the examination usually comes down to the examiner's impression on novelty and nonobviousness. Each claim of the patent is rated as accepted, objected to, or rejected.

Rejections and Arguments

In the event your patent application is rejected, your attorney or agent has an opportunity to respond to this initial office action with arguments and evidence to try to convince the examiner that he or she is mistaken. In addition, the response can include an offer to amend or change the claims in some way to get around the objections and rejections.

Patents Expert Pointer

To win a patent in the examination process, you have to be the champion of your patent, an optimist, and a fighter. When a rejection letter appears, everyone on the team can become depressed. The patent expert approaches the rejection with the idea that the examiner must have gotten it wrong. Rethink your invention, step through the logic of the rejection, and figure out the weak point of the attack. It is the patent examiner's job to point out the weaknesses in a patent application, but it is the inventor's job to point out the strengths.

The examiner then reviews these responses and issues a second office action. This action should show that the examiner has considered your arguments and changes, and it will again include a list of claims that are accepted, objected to, or rejected. Again, your attorney or agent gets to file a response.

Typically, this leads to a third and final office action, which may be stated as a final rejection. It is certainly alarming to the inventor to see the heading of this letter. However, it turns out that all hope is not lost—a final rejection is not really final.

At this point, your attorney or agent can make a last-ditch effort to convince the examiner to change his mind. This often involves a phone call between the attorney and the examiner. Surprisingly, this direct communication is often extremely effective in reaching an agreement about a few more amendments. Suddenly, the final rejection gets replaced by an allowance, and bingo, you have a granted patent.

Appeals and Continuations

In some cases, it does not go so well. The examiner sticks to his position and your final rejection is indeed a final rejection. All is not lost. There are two routes forward:

1. File a continuation application: You can't change the contents of the written description, but you can effectively restart the examination process by filing a new set of claims that may do a better job of meeting the examiner's objections.

2. Challenge the examiner head-on: Objecting to the examiner's decision and rationale usually means that the record of the examination is put before the Patent Trial and Appeal Board (PTAB), which is a creation of the 2011 AIA and replaces the Board of Patent Appeals and Interferences (BPAI). This is a panel within the USPTO that includes the director of the patent office and the commissioner for patents. The PTAB will review the record as it stands, and make a decision as to whether it thinks the patent examiner got it right.

AIA: Change of Rule

The old Board of Patent Appeals and Interferences (BPAI) became the new Patent Trial and Appeal Board (PTAB). It is no longer possible to challenge another patent based on who invented first (the old interference system). Now these challenges are made before the PTAB for derivation—an attempt to show that another patent application filed before yours was actually derived from or copied from your own invention. This board, despite the new name, still hears appeals against rejections issued by the examiners.

If the PTAB agrees with the examiner and leaves the rejection in place, you then have two choices for going forward:

1. You can appeal further to the Court of Appeals for the Federal Circuit (CAFC). This is really the top patent appeal court and is the same court that handles appeals for the federal district courts from patent infringement law suits.

2. File suit against the director of the patent office in the United States District Court for the District of Columbia. The main reason some applicants choose this route is that it allows the applicants to bring in new evidence and new arguments that weren't part of the original record. Both the PTAB and the CAFC look only at the exact information that was available to the examiner, and simply try to decide if the examiner made the right decision. The district court allows for a fresh attack on the decision using new information.

However, the decision in the district court could be appealed by the inventor or by the USPTO, which lands it in the CAFC. In this case, the CAFC considers the new evidence. In a final twist, though, the CAFC gives deference to the original USPTO decision and tends to be suspicious of new evidence that you presented to the district court—after all, if it was all so clear, why wasn't the extra evidence presented during the original patent examination in the first place.

Restrictions and Divisional Inventions

One more issue that can arise during examination is that the examiner may decide your patent application really covers two or more inventions; that is, it lacks unity.

If this happens, the applicant has a choice of which part of the invention to pursue. Once this first part is fully examined and the patent granted, the applicant can then start a new process for the second invention (and so on if the examiner found there were three or more inventions covered in the original application). Each new invention originating from the original application is called a divisional.

Eventually, you could end up with more than one patent. Each of them would have identically the same written specifications and identical drawings, but the claims would be different.

For the patent included in the appendix, the USPTO decided it was a single invention, but the European Patent Office decided it was three separate inventions. In that case, the state-based agency that was paying for the prosecution and examination of the patent (the Washington Research Foundation) decided to pay for just one of the three divisional inventions and let the other two elapse without filing.

In hindsight, it chose wrong. Over the following twenty years, each of the two inventions it ignored grew into multibillion dollar inventions, while the one it chose to progress to grant never developed much value. However, all three components were granted as a single patent in the United States, Australia, Canada, and Japan.

One attraction of packing multiple inventions into a single application is that you can get the search on all of them in the PCT process by paying just one fee. Until 1995, this also provided a method to get a longer patent life. Until that time, the life of a US patent before expiration was seventeen years from the date the patent was issued. One could therefore pack four inventions into one application and all would get the same priority date.

If filed in the United States only, the inventor could choose to keep the patent secret and unpublished. The examination of a divisional application could be deferred until one of the other divisional applications was granted.

Each exam and grant could take three years. In this way, the fourth patent might not be granted until twelve years after filing, and the expiration would be seventeen years later, or twenty-nine years after filing.

This kind of patent was also referred to as a submarine patent because it was possible to keep its existence secret for many years, and then use it to sue for infringement by competitors who didn't really have a way to know they were infringing. There are protections for a business that innocently infringes a secret ungranted patent.

In any case, since 1995, all patents have a life of twenty years from the original filing date. There is therefore less incentive for inventors to try to pack multiple inventions into one application.

CHAPTER 13

The Costs of Patents and Where to Get Help

In This Chapter

➤ The expenses of getting a patent

➤ Expense vs capability in a patent professional

➤ Getting financial help for filing a patent application

A patent is very expensive to obtain. There are ways in which you can publish an academic paper or a book wherein the writer has very little or even no expense. This is not the case with patents.

What Are the Costs?

Expenses just for the government filing fees for the application are considerable. However, the major expense is for your attorney or agent. Those expenses are paid in two parts:

1. Part 1: Preparing the application

2. Part 2: Managing the examination process and getting your patent granted

Filing Costs

At the very outset, the expenses are low. If you are an individual or small business that qualifies for micro entity status under the 2011 AIA (see Chapter 10), then your initial fees for filing a provisional application will be under $100.

From that point, the various fees are listed on the patent office's website: www.uspto.gov/ patents/process/index.jsp (click on View Fee Schedule on the left of the page). When you file the main application, there is a filing fee, but the amount of the fee increases based on the number of claims in your patent application. You pay extra if you file on paper rather than electronically. Then there is a fee for the patent search and for the examination. Most of these are reduced to a total of around $2,000 in total for the small business entity unless you want expedited examination, which will more than double this amount.

If the patent is granted, there are additional fees at the time of grant, and later on there are maintenance fees to keep the patent in force, which adds several thousand more in expenses.

Fees for Preparing Your Application

The bulk of your expenses will go to your patent agent or patent attorney for helping you write (draft) the patent applications and claims, and then shepherding the application through the patent office examination process.

Patent agents are usually used by individuals and small companies who are under great pressure to limit expenses. Overall, the sheer hourly cost of the work done by a patent agent will be about half the price of having the same work done by a patent attorney.

Patent Vocab

Drafting, filing, and prosecuting are the three important phases of professional work that you need to pay for to get a patent. *Drafting* is the term for writing a patent application. *Filing* is the process of putting together the appropriate forms and documents to get your application under way in the patent office. *Prosecuting* describes the process of shepherding the application through the patent examination process toward getting it granted by the patent office. It does not include any part of litigation over infringement.

In European countries and many other countries around the world, the vast majority of professionals who draft and prosecute patents through the patent office exam process are patent agents. It is only the United States where more expensive patent attorneys predominate.

To become a patent agent, a person needs a bachelor's degree in a relevant area of technology such as biology, computer science, pharmacology, physics, or some area of engineering. Then that person takes a preparatory course in patent law and then takes the patent bar exam.

A patent attorney usually also has a technical degree, but also went through law school to become an attorney, then passed the bar exam, and then went on to take and pass the patent bar exam.

There is no simple answer to what type of representation you should have for filing and prosecuting your patent application: self-representation, agent representation, or attorney representation. Sometimes, the sheer inability to muster the necessary funds rules out an agent or an attorney. On the other hand, just because you have the funds to hire an attorney doesn't mean you will get excellent work done for you.

If you represent yourself, there are numerous fatal mistakes you can make. However, it is possible you will be able to get a good patent granted with no more outlay of expense than what the patent office charges for filing fees. To succeed, you will have to do a great deal of studying before you file. You can also get copies of multiple patents in your field and use the style and structure of these other patents as a guide.

Patent agents can do an excellent job, although they may not have the full perspective that goes with participation in litigation of granted patents. Patent agents are never involved in patent infringement litigation and cannot help you with the contracts you may need for assigning or licensing your patent.

Patent Attorney Horror Story

Just because you manage to hire a top-level, expert patent attorney, does not mean you will get an excellent result. I think most attorneys and law firms are highly reputable, but it may be helpful to share a horror story from my own patent filing experience.

Inventor Beware

Just because you hire a famous top-level, expensive law firm doesn't mean you will get excellent, or even competent, help. No matter how much money you pay, it may seem like chicken feed to a big firm. Also, you just are not worth that much to a firm. A company like Exxon will spend tens of millions of dollars with a law firm. How much is the business with you going to be worth in comparison?

I was not at all new to the patent application process and had filed a variety of successful patent applications through various firms, but now I had my own small startup company and wanted to hire an excellent attorney from an excellent firm to write up the application for a new invention.

After extensive online searching, I had identified the top patent attorney at one of the largest, most highly regarded, giant business law firms in the United States. She was based in its Palo Alto, California, office. I had a preliminary discussion, determined I could handle the projected fees, and then I scheduled an initial in-person, one-hour meeting to discuss the patent.

When I arrived, there were seven patent attorneys in the room. They explained that after reviewing the initial information I provided, each had relevant input. I launched into my presentation and left time for questions. Everything seemed great, but the bill for the one-hour meeting included all seven attorneys billing at $500 per hour. The initial one-hour meeting cost me $3,500. It later turned out that none of these attorneys ever worked on the patent.

I had a few more requests for information, but I had filed a provisional and there was a twelve-month time limit. As the date when the definitive application must be filed approached, I became increasingly anxious because there was no draft application available yet for me to review. I tried some telephone meetings to make sure the lead attorney understood everything.

Bills for research and discussions and planning and consultation and drafting came in steadily, running up over another $20,000. However, still no draft application was available for review, and there were just three weeks left to get the application filed.

At this point, I started to lose patience and became concerned. I wouldn't have time to review the application, get a revision, and reassess. I insisted that I must have the draft by two weeks before the filing deadline, but no draft was available. With just two weeks left and tens of thousands spent, I was reluctant to start from scratch with a new attorney or agent. I was sure that with a reputable firm and a prominent attorney, I just needed to be insistent and patient for a little while longer. At one week, despite several e-mail requests, there was still no draft application.

The filing was due on a Wednesday and I insisted I must have the draft by the Saturday before, so I could have Sunday to review it. No draft document appeared. On Monday, the patent attorney called to say nothing had been done yet. She explained that they had assigned the application to a new first-year attorney, fresh out of law school. Each of my requests had been passed along to the new attorney, and he kept assuring the senior attorneys that he was nearly done. There would be another $10,000 in bills, but unfortunately, he had just informed them that he had quit the firm to start his own law firm in general law with a family member. He had just revealed that he had not done anything at all.

With just over forty-eight hours left to file, the law firm found an attorney in its Texas office who was familiar with the field of my patent, and he stayed up for nearly forty-eight hours writing it from scratch. He filed it in Hawaii to get a few extra hours before the end of the filing deadline.

It was a mess. There were numerous missing parts. It misstated the invention in several ways, and could not be repaired. The patent was rejected by the patent office.

The large, famous law firm terminated my selected attorney, then sent a letter saying it wanted another $100,000 in fees with no particular evidence of the work done to support the billing. The bill was later dropped when the management was confronted with a detailed summary of events. I was relieved when that happened, because I was not excited about personally suing one of the largest law firms in the United States for malpractice.

OK, nothing that bad is going to happen to you. However, there are several important lessons to be learned.

You should expect your lawyer to start early and promptly produce an excellent product. If that does not happen, then move on quickly. Your number-one concern is getting an excellent application. Another time, when I had a patent agent take my initial provisional, talk with me for an hour, then redraft the application in the appendix, the job was done in a few days and much of the work has held up excellently across many years and many challenges.

One measure you can take, if you can afford it, is to have the patent drafted initially by a patent agent. Then have it redrafted by a patent attorney. The logic here is that with an initial draft, everything will be put in order and your communication with the patent attorney will be more efficient.

Patent attorneys and patent agents bill by the hour, so a patent attorney is less expensive if you have everything prepackaged in a way that the attorney can readily understand and work with. This approach also has the advantage of having two separate experts look at your invention so that the "two heads are better than one" principal has a chance of leading to the identification of weaknesses that were missed in the first pass during a stage when they are easy to correct.

As soon as you have an initial professional draft from a patent agent, you are ready to file no matter what else happens. If the money comes along that allows you to do more, then you may get an even better package assembled. If not, you are still ready to go.

It is worth pointing out that I have also been pleasantly surprised when an attorney quickly and inexpensively turns out an excellent application that sails through the patent office with allowance at the first office action.

Patents Expert Pointer

A good way to choose an attorney or agent is to conduct a quick interview on the phone. Some will even provide a reliable estimate if you can confidentially outline the nature of your invention.

The Bottom Line

The attorney or agent cost for preparing, filing, and prosecuting through to grant of the patent still varies greatly. A general estimate of the typical expense for a simple mechanical invention is about $5,000 when using a patent agent. At the other extreme, the very long, complex patent shown in the appendix ultimately cost nearly $500,000 for drafting, filing, and prosecuting through to grant in the United States, England, France, Germany, Austria, Canada, Australia, and Japan. These probably represent the two extremes. You can consider this in comparison with the $4 million in litigation expenses in court actions against two infringers.

When you are looking at expenses from the front end—you have an invention and no money and you want to get it filed—then $5,000 in attorney expenses and $4,000 in patent office expenses may sound astronomical and overwhelming. Many years later, if your invention is a commercial success and you have to deal with infringers, then corners cut in the early days may cost you dearly as you try to compensate for them on the back end.

Getting Help to Pay for the Patent Filing

One bit of good news is that if you have a great new idea, there is a decent chance that you will be able to find help for the costs of filing and winning the grant of the patent. The challenge is in determining what you should give up in order to get the help you will need.

Investor in an Idea

You will quickly learn that it is a much easier task to get investors in a granted patent than it is to get investors in an early stage invention. It's not that anyone is trying to be mean to you; it is just that the uncertainty level is greatly increased. Often you will need to give up more in order to get help because the risk is higher for the investor.

The amount of money you are looking for may be relatively small compared to the kinds of investment needed for growing a company or developing a product launch. However, that may not help because an investor is often concerned with how much research and study will be required to make an investment decision, and the time may be harder to justify for a small investment.

Patents Expert Pointer

Most technology investors are looking for new technology that is already covered by a granted patent. However, if you take the time to search carefully, you may find that some large companies in your field, and some specialist investors, are out there looking for inventions at a very early stage. These people know that if they look at an idea and believe it will turn into a granted patent, they can get great value for money by coming in early.

You have to impose strict confidentiality to assure that your efforts to find supporting funds don't end up torpedoing your invention through unintentional disclosures. You also have to avoid spending hundreds of hours pitching your idea to dozens of investors who weren't really that interested and have depressingly uninformed negative ideas about what you have done.

However, if you reach the right person, the benefits can be large. The best situation is where you find an investor who has dealt with early stage ideas like yours before. Often, such an investor will know patent attorneys to refer you to. For the patent attorney, your invention will then represent part of a stream of business from that investor so you will get much better attention.

Your Employer Pays

In the very earliest stages when you are realizing that your employer, whether a private company or university, will be taking ownership of your invention, accepting this loss of control is very painful. However, once you come face-to-face with the expenses, you will quickly realize the huge upside of assigning the invention: your employer will be paying the bills for the patent application filing and prosecution.

Yes, you will need to pitch the value of your invention to your employer, but if he believes in the potential value, then you may get top-of-the-line support. Once again, you may benefit if your employer uses the same patent firm for a stream of patent work. This means your project will get more attention and more effort.

Later on, once the patent is granted, you may still have a chance to go out and put together a startup company, get potential investors lined up, and then license your own patent back from your employer. You may have to pay back the patenting costs and give a share of your future earnings back to the university or company, but this pushes the expense into a time frame where you are in a much better position to handle it.

Developing a Relationship with an Interested Company

One other route to success is to approach the scientific or intellectual property leaders at a large corporation. This is complex because you have to take great care to protect confidentiality on both sides. You will be concerned that the large corporation will take away your idea and run with it. The large corporation will be concerned that you have invented something that it has already developed and that later, when it brings its product to market, you will falsely accuse it of stealing your idea.

Another obstacle to bringing in outside inventions is a "not invented here" mentality that places a higher value on the corporation's own science. You just won't have the champions inside the company like an in-house inventor will have.

Nonetheless, this sort of thing works best when you have an idea that is a major advance in a field somewhat near the company's focus, but is so unique that it is unlikely anyone in the company has thought of it. This can work with small changes and small advances, but these are often less attractive to the corporation.

When everything works out well, a company representative will agree to a confidential evaluation. If he is interested, he may assign the company's own high-cost superstar patent firm to write and file the application with what seems to you like an unlimited expense account. The company's idea is that it wants to license the patent (and it will require you to give it a first right to license before you can offer it to anyone else), and if it does so, it wants the best possible patent.

PART FOUR

Using a Patent in Business

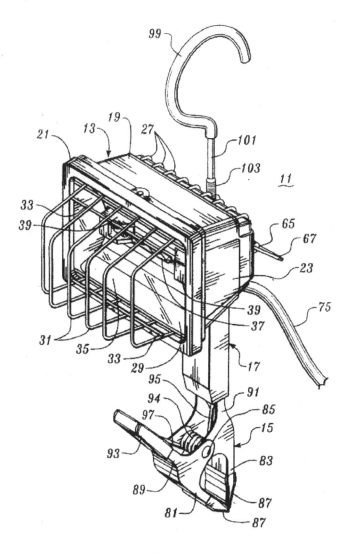

FIG. 1.

CHAPTER 14

Developing a Patent Once It's Been Granted

In This Chapter

➤ The patent as intellectual property

➤ Turning your patent over to someone else

➤ The importance of timing

You've succeed in getting a granted patent, now you want to know how this brings financial reward. The answer is that getting the patent is the easy part.

This goes counter to the vague notion held by some first-time inventors that getting a patent is like having your rock-and-roll song hit number one on the charts. You have to keep in mind that there are about 350,000 patent applications filed every year in the United States, and about 240,000 of these get granted. There have been over four million patents granted in the United States in the past thirty years alone.

The Value of a Patent

The good news is that a patent lends considerable strength to a business idea just as it lends some durability across time. You can launch a business based on a new idea without a patent, and then make a race for market share. However, if the product is successful, you will rapidly be faced with numerous competitors offering the same thing that you have developed but at a lower price.

At this point in the process, it is helpful to start thinking of your patent as a piece of property (intellectual property) instead of thinking of it as a device or invention. Like any piece of real estate or any item of personal property, it can be bought or sold, and therefore there is an interest in trying to predict what its value or price should be.

Because patents are generally concerned with useful commercial processes, the value of your patent can be determined in commercial terms. That value will result in part from the price and profitability of the product produced under the patent and in part by the market share you can get access to. Your own share in the value of the patent varies greatly depending on the business arrangement under which it is developed.

As you will see, there are several ways to show that your patent is valuable, but these ways differ greatly in terms of the financial value to the inventor.

At one end of the range, you may be an academic inventor in a university where there is some potential financial value to your department, but where the actual inventor is determined in advance by university policy to get absolutely nothing.

At the other end of the range, you may be able to build a successful startup company around the patent. You line yourself up to hold a major portion of the shares in the company, your invention takes off like a rocket, the company goes public in a huge IPO (initial public offering), and you retire a billionaire at age twenty-three.

In these two scenarios, the inventor who gets nothing may have invented a major new medical imaging system that will save millions of lives and sell for $10 million a piece to every hospital in the world, of which the inventor gets nothing. The twenty-three-year-old billionaire may have invented a new way of delivering dog food that forms the basis of a company that goes out of business completely bankrupt a year after the initial public offering, but the inventor has already cashed out and is already building his mansion in the Bahamas unaffected by what he has left behind.

Inventor Beware

There are companies advertising on TV that if you give them your idea, they will take care of everything. But have you heard about any successes that started this way? This may sound like a promotion for snake oil, because probably in a way, it is. Don't fall into this trap.

Letting Your Employer Do the Heavy Lifting

In the previous chapter, it is pointed out that you may have no access to the kind of money it takes to actually get a patent filed and granted. It is all good fun to imagine being a millionaire someday or to be able to tell everyone how much you believe your invention is worth, but for most solo first-time inventors trying to get a patent, the prospect of putting together $10,000 over two years' time for something that may never amount to anything is overwhelming.

Investors who will put money into an early stage idea for which no patent has even been filed are rare. If you do find someone willing to take the risk, you may discover you have to give up 80 percent of the control for that $10,000, when you believe that two years from now, once the patent is granted and it has gotten some great press coverage, that $10,000 would buy the investor less than 1 percent of the control.

One common way out of this dilemma is to let your company or university pay the patenting costs and develop the commercial use of your invention. This can bring an inventor some significant income if he works somewhere that provides the inventor with a share of any income eventually received.

When an invention is close to a business's main commercial focus, this is an excellent way to turn an idea into income. Rather than get personally involved in developing the business, you return to the lab and make your next invention—and so on.

However, if your company's business focus is not squarely aligned with the potential markets for your invention, you may be frustrated to learn that although your patent is owned by your business and although it spent the time and money to get the patent filed and granted, it doesn't have an interest in actually developing the business potential.

Sometimes a company will hold on to a patent as a potentially valuable piece of property, waiting for a future opportunity to arise when your patent can be bartered or sold.

Even worse, your company may have a hated competitor that overlaps and competes in one area of your company's business. The other company can be blocked from its efforts to do business in the additional area of your invention—your company threatens to sue for infringement if the competitor enters the new area. Neither company actually does anything in the area as a result. Your patent is used to obstruct and threaten but never generates any income.

Patent Vocab

Nonuse occurs when a privilege, right, or claim has not been exercised. One of the most frustrating scenarios for an inventor is to watch his patent fall into the hands of an entity that will not see it to fruition.

One other very common trap of nonuse arises with universities and government employers who simply don't do any business in any area. They may be required by policy or by law to retain ownership of your patent but have absolutely no motivation to develop it or sell it. In a related scenario, your university employer has an Office of Technology Transfer, which has a list of 147 patents granted to its employees that it posts on the Internet as available for licensing.

The idea here is that academics may make inventions, but your university believes that academic scientists should be doing basic research and not doing business. However, the university or government lab has some desire to provide benefits to local industry by providing its inventions for commercial use. The problem is that no one has any passion for developing your invention and companies may not like the idea of licensing.

It is true that every now and then an energetic local company with a business in the general subject area of your invention may decide to look through the list of available technologies and stumble on the listing showing your invention. If it works out terms to license or gain control of your patent, then your invention could get turned into a valuable commercial force in the marketplace. Depending on the arrangement under which the company gets access to your patent, this could bring some income to you (or to your department).

In some cases you, the inventor, will lose patience waiting for some existing company to come along and offer to develop the invention. You could try to promote some activity by contacting businesses that you believe could be interested, and drawing their attention to the patent. One other option is the American classic situation where you quit your job, leave the government or university, start your own new company, and license or buy the patent from your former employer. In this scenario, you make your own opportunity—you become the entrepreneur.

Letting Someone Else Turn Your Invention into a Business

What are the important factors that affect the likelihood of an idea becoming an invention, an invention becoming a patent, and a patent becoming a successful business? There are many ways in which this sequence occurs, but the patent system creates an opportunity to do the business equivalent of what the ancient inventor Archimedes once said of his levers: "Give me a place to stand and I will move the world."

No One Cares About Your Invention More Than You

The main problem with letting anyone else develop your invention is the simple immutable fact that no one cares about your invention the way you do. My father was an inventor and he put this idea to me in exactly these terms: your invention is like your own child, and no one cares about its success more than you do. Not only that, but it is actually your responsibility to do everything you can to help it succeed.

Giving a Monopoly to the Inventor is the Essence of Patents

This idea was well understood in the seventeenth-century England of King James I, when parliament applied the 1623 Statute of Monopolies to aggressively force the king to stop giving monopolies to political supporters, family, and friends. The only monopolies to be created under English law from that time forward would be monopolies provided directly to inventors who were the true first inventors of their own technological advances.

England had grown from a weak agrarian country to become a strong commercial force in the world in the three centuries since Edward the III started giving monopolies to inventors in the 1300s (see Chapter 2). The further refinement put in force by Parliament in 1623 helped drive England's march to become the dominant political, military, and technological country of the world in the following 150 years.

When the United States became independent in 1776, it enshrined the English idea on patents and monopolies as an important element of the US Constitution.

During the three years when Thomas Jefferson started the US patent system (1790–1793), he strengthened it by introducing the idea of patent examination to make sure that among those who claimed to be inventors, only true inventors of significant advances would get these monopolies. What kings, parliamentarians, early US presidents, and American founders all understood was that if you gave inventors a chance to protect and develop inventions, the entire nation would benefit and lead the world.

Even today, countries with no history of a vigorous patent systems have been followers who get their technologies second hand from the small number of countries that are the engines of invention in the world—the UK, the US, Japan, and Germany (since it developed its patent system in the late 1800s and expanded it into the European system in the 1970s).

In one recent survey of the important inventions from around the world from the past three centuries, 54 percent came from the UK (mostly Scotland and England), 25 percent from the US, 5 percent from Japan, and about 1 percent from the rest of the world combined. Within the United States, California generates five times more patents per year than any other US state.

At the heart of all this is the British idea that inventors should be given a monopoly to develop their own inventions. The command systems of the Communist era failed to compete. Until recently, modern industrial powers such as China, South Korea, and Germany have relied on technology licensed or copied from the countries that generate innovations. However, the growth of entrepreneurship in China, South Korea, Russia, and India together with the willingness of those national governments to support innovation and promote business development is steadily changing the technological geography of the world.

At one end of the two extremes for patent generation is the solo inventor entrepreneur model whose inventions transform the industry, what might be called the California model. At the other extreme is the Korean model with large corporations generating tens of thousands of patent applications per year and a patent office that rushes them through to grant five times faster than the USPTO. The distinction is captured nicely by considering Apple versus Samsung. If you're reading this book, you're probably more in the zone of the California model.

The Timing of Inventions

An important element to consider when deciding on the amount of personal risk you're willing to take to launch your new technology business is timing.

Some inventions result in a product that can be rapidly put into production and will almost immediately experience a high level of interest by consumers.

At the other extreme are inventions that define a significant change in technology. These often become useful only when their potential customers come to understand the new technology and have time to change their way of working or acting in life. Such inventions can take twenty years to develop their market. In this latter situation, the patent may expire before the product becomes a commercial success.

Inventions of high, immediate value tend to be incremental improvements in an existing product in an industry with rapid progress to market.

Consider the opposite extreme with a new type of pharmaceutical—ten to fifteen years of development, testing, and clinical trials will be required before the company can begin to market and introduce it.

Patents Expert Pointer

Some pharmaceuticals take so long to get from the laboratory to store shelves that they outlive their patent while still in the developmental stage. For this reason, the USPTO grants patent life extensions for some pharmaceuticals.

In addition to slow testing and regulatory processes, other inventions simply require a long time for users to adjust to. Still other inventions represent a major advance but need additional refinements, improvements in manufacturing, or a larger scale of production to get it into a price range where consumers will be interested.

To the inventor, the usefulness of the new product may seem obvious. However, finding out how to actually get the public to understand and desire the product is far more difficult.

These are some of the reasons the vast majority of inventions are either simple mechanical inventions or improvements that emerge from within existing businesses that already have extensive relationships with consumers.

Simple mechanical inventions have the following qualities:

➤ Easy to manufacture

➤ Rapidly brought to market

➤ Easy for consumers to understand

➤ Inexpensive to modify existing manufacturing methods

➤ No government regulatory process

When an invention is complex, difficult to manufacture, creates an entirely new market because nothing like it exists, then the initial costs are large and success may take many years.

If you can sustain the cost of development and tolerate the many years of slow growth, then you will be fine if your invention becomes a mammoth success eighteen years after the patent filing. All your efforts in the previous two decades will pay off, and you will have a powerful presence in the market to continue competing once the patent expires.

If success occurs in twenty-two years, others will immediately rush in and all of your hard work in development and marketing will mostly benefit your competitors.

Inventor Beware

A complex and radically innovative new invention is exciting, but it may take many years to develop into a successful business. Think very carefully about how you will develop a successful business based on your invention before you sink a large amount of personal funds in the patenting process

Although simple inventions may be more easily brought to market, they may also be easier for a competitor to copy or to invent around. If a patent does not mark out a large area of novelty, then it may be possible for a competitor to see ways of building something similar that takes advantage of the inventor's inspiration and market development but has enough changes to evade the patent.

Another problem with inventions that reflect a small step or change in an existing type of product is that many others will be able to make their own small changes and improvements independently. Other patents may be issued in your field for inventions that provide the consumer with similar benefits and improvements for an existing type of product, but improve a different aspect of the product than the one you were focusing on. For example, you develop a dramatically better type of rearview mirror for cars, but consumers are interested in video-based backup systems.

Similarly, an invention that improves fuel efficiency of cars may prove to be of much greater competitive value for selling cars than an invention that improves the rearview mirror. Here you have definitely improved a product, but so many others are involved in improving other aspects of the underlying product (cars) that your simple, fun, inexpensive, and extremely original innovation simply can't compete with the market power of other inventions in your field.

There is no certain way of determining in advance if your invention will be ahead of its time (so advanced that consumers can't really accept it until after the patent expires), at its moment (ready to go and readily accepted early in the patent life), or left in the wake (swamped by the rapid progress of other inventors in your field).

You have to carefully consider the elements of time to market, degree of marketing/ advertising needed, and the extent in which consumers will need to change their behavior for them to be interested in your invention.

This is the area where Steve Jobs excelled—developing products that provide a significant change (the computer mouse, type fonts on screen, laser printer, iPod, iPhone, iPad), and using his own judgment and intuition (rather than focus groups and marketing surveys) to successfully alter the way that vast numbers of people work, live, and express themselves. He also understood the use of design (appearance), manufacturing excellence, and effective advertising to achieve Apple's commercial goals by steadily expanding the size of its customer base.

 Building a Company Around Your Patent

Like pioneers and adventurers through the ages, the inventor-entrepreneur still captures the human imagination. To succeed, you need the perfect mix of creativity, discipline, foresight, knowledge, charisma, leadership ability, and competence. However, you won't really know if you have all those characteristics until after you succeed. This is a much broader challenge than simply making an invention and getting a patent.

The First Steps to Building a Startup Company

In the "California" model of success, a solo inventor gets money from family and friends to make a prototype, attracts venture capital into a new startup, rushes to market, grows the business rapidly, and then cashes out with an initial public offering of stock—or goes on to develop new related products to grow the business. Variations on this model take place all over the world and are a major engine of human advance.

Samsung is now a giant that produces a huge number of new technologies that have a great impact on many consumers. It started when a young man from a wealthy family launched a grocery store and began manufacturing noodles. With some commercial success, the business went into making wool products. Its success in that business led to offering financial services such as insurance. It then expanded in all directions with numerous subsidiaries until eventually it was involved in dozens of industries all at enormous scale with thousands of people employed just to make inventions.

Starting a Tech Company

Starting a new business is actually simple, and you can find many good guides in books and online. In essence, you just need to file the appropriate paperwork to create a business.

You can start as a sole proprietor, which is you doing business as an individual. More commonly an inventor will choose to set up a new corporation because this allows for investors to get shares when they provide money.

Once you've made a few basic choices about corporate structure (a C corporation is best for bringing in various investors; an S corporation may be better if you expect that only a small number of friends or associates will ever be involved) and filed the papers, you have created a new corporation.

The invention you made will either belong to you, the inventor, or to a university or employer. If you own it, you will need to assign it to your corporation so that it is the property of the new business. No one will invest in your company if the corporation doesn't have the asset—control the rights to the patent, if granted, or owns the granted patent itself.

Many inventors choose to start a company with a few other people who can bring in necessary experience. You might invite an accountant, some other associates in your field of technology to help with product development and manufacturing, a potential key investor, and perhaps someone with experience at managing and running a startup company. If you line up all these people in advance, you can all sign a preincorporation agreement that explains what the business will try to do and what everyone's role will be.

You can create the company with a chosen number of shares of stock and give everyone a significant percentage of the shares as of the creation date of the company. In this way—although you don't have money to pay these people and although they don't have to invest anything—you provide a potential reward to each of them, while taking advantage of the collective experience of the group.

Pitching the Owner of Your Patent

If your university owns your invention, you will need to make a presentation to key university people. You will be trying to convince them that rather than simply listing your invention in a brochure of available technologies for industry licensing, they should effectively give the invention back to you, the inventor. You will be seeking an assignment of the patent (transfer of ownership) or an exclusive license (total control of the invention). In exchange, you offer to pay the university royalties as the invention becomes a success. You may also be able to offer the business or university shares in your own new company.

When the university, the business, or a sophisticated investor looks at your opportunity, he will be looking at the technology you have invented, at you as the prime mover, and at the team you have assembled. Even if the technology looks great, if your team is not convincing as a group capable of starting and running a successful business, then it makes no sense for the patent owner to assign or license to your company and it makes no sense for the investor to risk losing money with you.

If you plan well, you will get a great team assembled, set up the corporation properly, get the patent assigned or exclusively licensed, and put together enough initial small investment money (from yourself and shareholders) to get the ball rolling. You will then need to correctly determine how much money will be needed and when it will be needed to make the new company and product a success.

Venture Capital vs Operating Business

There is a widely held misunderstanding among potential entrepreneurs that getting an investment from a venture capital company is the number one measure of success. In fact, venture capital investments can be very destructive to the inventor. In one company I started, my original partner and I put in twelve inventions over ten years, went through several rounds of investment, had successful products, and developed partnerships with major manufacturing corporations. The company was eventually sold for about $40 million and I got $100 out of the whole thing. The employees and scientists got nothing at all.

You need to keep in mind that investors put their money into your business for one reason. Their objective is not to help you personally. Their objective is not to help bring a product to market. What they want is to put some money into a transaction and get a great deal more money back at some reasonably near time in the future. They don't really care what happens to you or your company.

It is your job to protect your own interests and your company's interests. Unfortunately, most venture capital companies have extensive experience dealing with hundreds of startups, and they have the power that comes from having the money. They also have the fear and the knowledge of what happens when it all goes bad: they lose a fortune, but the entrepreneur just walks away and starts another business (from their perspective).

How Do Venture Capitalists Work?

Venture capitalists are people who are investing other people's money. The money comes from various wealthy individuals or from pension funds or even from banks.

Investors put money in bonds, CDs, and saving accounts to get a slow accumulation of money from interest at a predictable rate with no risk of losing the original starting capital. They go to the stock market to try to get gains of 10 percent or 20 percent in a year of clever trading with a risk of losses of similar scale. They can also go to a VC fund that boasts that it once invested $1 million in a startup and sold out for $100 million in five years. The investor could lose everything, but he could also get a 10,000 percent increase in the value of his money.

VC companies assemble experts such as past CEOs, bankers, and experienced, successful tech investors. They try to convince investors that if they put their money into the VC company, the experts there will study a thousand startups to find the one diamond in the heap of coal.

The investors have no idea what startup companies the VC firm will look at or what industries they will be in. They really don't care about the inventors who developed the startup companies. They just want to be sure they won't lose their money and that their investment will do well.

Sure, later on, when the VC has chosen ten companies to invest in, it may all look different. Six may have gone bankrupt after the investment. Two may still be running but increasing in value at just 5 percent per year. However, one of the investments may become a huge success (hopefully this is you).

The VC company will boast to its investors about its great wisdom in picking a winner. The investors will become instant experts about every detail of your company that they have cleverly put their money into. They will be telling everyone who will listen about what a genius the inventor/founder is and how they were so clever to trust and invest in you. However, when and if the time comes to shut it all down and get the money back for their investment, they won't be bothered if nothing is left for the original startup team.

Avoid Dilution

What you have to do is to try to limit dilution of ownership and control at each round of investment. Consider the description at the beginning of this chapter about the solo inventor who is out of a job and has to give up 80 percent control at the outset to get the first $10,000 to file his patent. In the next round, the company raises $500,000 to do development, set up manufacturing, and start marketing—for this they sell 60% of the company.

The first round left him with 20 percent ownership, the next round took 60 percent of that away, so now he owns just 8 percent of the shares. Orders start coming in and they need $3 million to get raw materials, scale up manufacturing, and hire a sales force. They get this by selling another 50 percent so he is down to a 4 percent share.

Inventor Beware

Dilution is the inventor's biggest enemy. At the beginning, your invention makes up the full value of the company. However, once the invention is assigned to your business, cash becomes king, and cash is something you probably have very little of (when compared to your investors).

You start off with a big share of your startup company because you brought all the value. However, your share gets smaller and smaller as round after round of additional investors join in to help make the business successful. This process is called dilution. The best way to beat this is to keep in mind from the beginning that you must have a large share of your business from the start, and you must constantly fight to keep your ownership percentage of your company as large as possible.

The business starts to grow, but demand is too high for you to meet. Bad press accumulates about poor delivery times, falling product quality, and customer dissatisfaction. Three large corporations rush to fill the void you are leaving by failing to keep up with demand.

An emergency board meeting is called to address the real fear on the part of investors that the company will collapse. They vote that all existing shareholders are allowed to triple their investment and triple their shareholdings if they invest within seven days. Suddenly $9 million pours into your company, and within two months you have solved all the problems in manufacturing and the ongoing profits are sufficient to stay in business. However, your share is now down to just over 1 percent.

The board is pleased with its success and the profits, but they want to buy some smaller companies to broaden the commercial base. They raise $20 million from other investors, giving up 30 percent ownership, and your ownership goes down to 0.66 percent.

Votes at the next board meeting, as always, are apportioned by the number of shares. The major investors vote to create a special class of stock available only to those owning at least 5 percent of the company.

These new shares will get preferential rights over any income from a sale of the company that effectively doubles the financial position of the special shareholders. It is small change for them—someone holding 40 percent now goes to 42.5 percent—but you go from 0.66 percent down to 0.33 percent.

When the sale of your company takes place, the board decides that everyone holding at least 5 percent of the shares will get a bonus out of the investment proceeds, shares in the new company, and part of the sale proceeds. Because of getting shares in the buying company, they agree to sell your company for one-third of its market value. You should be getting a picture here of how unprofitable success can be if you bring the idea but everyone else brings the money.

Why will VCs be so harsh and uncaring? Well, any investor can choose among numerous venture capital funds. Investors will go for the company that can show it is better at squeezing the last penny of return out of its investment when compared to competing VC firms. This is why they are sometimes referred to as vulture capitalists.

Are there any real alternatives? Banks are hard to deal with because they prefer to invest in safe, established companies with small but predictable growth. This could be a new bakery in a community where the nearest bakery is 10 miles away.

The person opening the bakery has two successful bakeries elsewhere in the city. Each of his bakeries has grown at the rate of 20 percent per year over the past five years. Marketing surveys show that a thousand people near the proposed site love baked goods and are frustrated by having to drive 10 miles to this baker's other store. The proposed site is next door to the post office and the elementary school. Across the street is a purist espresso coffee shop that makes excellent coffee but doesn't sell any pastries. You can probably get the picture of what sort of low-risk business investments a bank likes to take on with a loan.

How about friends and relatives? Well, that sounds good, but what if you fail and lead your Aunt Rosie to lose her entire life savings of $10,000 just after she retires?

The best possible way out of this difficult situation—if you can do it—is to figure out how to actually earn money with products that you sell. Fortunately, in the age of the Internet, you can open a store on Amazon.com and have marketing, customer access, sales transaction capability, even shipping almost immediately. If you can just make a good saleable product, then you will have income. The more income, the better you can do at meeting demand.

If it all goes well, you may never need investment—or at least you may be able to get by with a low-risk bank loan once expansion is well under way. Loans may seem worrisome from some points of view, but you don't give up ownership. All you have to do is pay off the principal and interest.

How to Describe Your Invention to Investors

Despite the cautions and warnings stated above, the fact is that you will probably need investment to develop your invention into a valuable product.

A friend recently told me a story of an internship he did many years ago with a wealthy inventor who had purchased a large collection of the invention models that inventors used to have to submit to the patent office. In cataloging the models, he found many wonderful machines from long ago that accomplished things that were great, but that no one had ever heard of. In fact, many never went beyond the model even though they met a significant commercial need, were clever, worked, and could have been huge successes.

These models—orphans of invention—are typical of many of the millions of granted patents. The invention really was great, but the inventor just didn't manage to get the product developed. The bright light of imaginative breakthrough receded into the darkness of this musty collection, never to reach the public.

There are probably various reasons this occurred. In some cases, the inventor may have been a great tinkerer but had no interest in business. In other cases, there may have been a will to make a business, but the person was just too disorganized to make it happen. In many cases, the inventor was not only creative, but was also potentially a good businessman, but he didn't have the marketing skill to be able to pitch the idea to investors to raise sufficient capital to make the business a success.

Although Chapter 10 describes how to write an invention disclosure, that document is probably far too technical to gain an investor's interest. When you have the opportunity to approach investors, you have to understand their mindset and what they are looking for.

Venture capital is a broad term. There are angel investors who put a smaller amount of money into early ideas, small funds looking for projects in special areas that haven't been snapped up by the majors, and large established firms with scores of employees and thirty years of famous successes.

How do you get to present to these people? Remember that their success depends on finding you; however, they are inundated with hundreds of opportunities.

The Fast Pitch Contest

One way to really learn your place in this ecosystem is to sign up for a fast pitch contest. This has nothing to do with baseball. You get lined up with dozens, or even hundreds, of other entrepreneurs, and you have two minutes to present your idea and your company in a way that will convince the investors in the audience that your opportunity is a standout.

Here you are with one of the great advances of the twenty-first century in your PowerPoint, and you know you could make these investors wealthy beyond their wildest dreams. But your pitch feels more like a painful off-key first round audition for *American Idol*.

If you give the best pitch, you get a prize. But you probably still don't get an investment. You see, this actually is just entertainment for investors. It is all justified as a learning experience for the new entrepreneur.

Cold Calls

Other more tolerable avenues of approach include making dozens of cold-calls to VC firms. This sometimes works. You have to have a one- or two-page executive summary ready to e-mail. You call through a list of VCs after first doing your homework. You list those firms whose website indicates that they make investments in your area of technology. Ideally, you identify the particular person in the firm who has experience in this area.

Now comes the cold call. You don't have much time. This is really like the fast pitch contest, only much less humiliating. You call the firm, tell the receptionist who you would like to speak to, and say upfront that you have a new proposal (ideally on a granted, or at least filed, patent so neither side has to worry too much about confidentiality). If you are put through, you'll have a minute or two to capture the investor's interest. What you are trying to accomplish is an invitation or permission to send an email with your executive summary.

If you can't reach the person despite several efforts, you might just go ahead with the email without an invitation. Note that if your executive summary is an email attachment, you may be foiled by email systems that block all attachments from unknown email addresses. A query email can get around this. You send a brief email that has the objective of getting the investor to ask to see your executive summary.

If you have a CEO in your company that has pitched a few companies before, then that person can probably make a few calls and set up a tour.

Ideally, you and your CEO have a series of meetings with investors and investor committees at a few VC firms, giving a thirty-minute PowerPoint presentation with all the bells and whistles. Your idea is in play, and you may fly home with several different competing offers to choose from.

The reason the CEO is so important is that investors invest in businesses and not in inventions. They know that no matter how great the idea is, success depends on execution. So having a businessperson in charge who they can develop confidence in is extremely persuasive.

How do you get an experienced CEO? Well, these people are constantly moving from one company to the next as startups go bankrupt or get bought. They may turn up at fast pitch contests to look for new projects. You may get introduced to a CEO by a friend or business associate or university technology transfer office.

Getting this right is obviously important for you and for the investors. However, if you don't have an experienced CEO, then it's all going to be on your shoulders to have some way to convince the investors that you are both a creative technology source and a capable businessperson. This is more difficult, but not impossible.

Investors are looking for an executive summary and a business plan that shows you have a marketable, patent-protected product, as well as evidence that you understand how to develop a business. Remember, you are going to see them because you believe you have done your part (made a great invention) and you want them to help you get rich. They are seeing you because they want to make money—they want to know what you can do for them, not what they can do for you.

A VC firm may boast that it has helped a hundred companies form and develop. However, these are not warm, friendly advisors trying to help you. When they make an investment in a company, it is usually provided in pieces (tranches) based on milestones. They tell you what you have to accomplish by a certain date. If you don't make the milestones on time, they may pull the plug—stop delivering money and let you abruptly go bankrupt.

Even worse, you may be meeting your milestones, but they have learned of another opportunity that seems more likely to succeed and more likely to be huge. Unfortunately, VC firms get overwhelmed by fads and paranoia—they hear about a new technology direction that other firms are pouring money into and your company doesn't have the cool factor of the month. They don't want to get left behind and miss out on the next big thing. You think you have met your milestone (progress in the lab, identifying a manufacturing facility, putting up a web site, hiring a marketing director, etc.), but they tell you that for various reasons they don't feel that you have properly met the milestone and so they are pulling out.

It is all very harsh, but somehow, many startup companies do succeed. You have to be nimble. You have to understand your investors, and you have to be a great salesperson.

Patents Expert Pointer

The patent expert knows that the best way to convert a great invention into a large amount of cash is to write a truly great and convincing business plan. Investments usually go into businesses, not into isolated inventions. Companies would much rather pay a high price to buy a company than to buy a patent that has not yet been applied to generate a successful business. The business plan is how you communicate to investors what you have invented and how it will drive a successful business.

The Elements of a Winning Business Plan

A business plan is a formal document that shows that you have worked through all the elements of putting together a successful business based on your intellectual property (IP). There are no absolute set rules for what you need to include.

There are many books on how to write a business plan. The example discussed here is a workable guideline. As indicated above, this is different from a description of the invention. In fact your invention description (disclosure) can be an appendix to the business plan, but the two are very different documents.

The Cover Letter

You start with a cover page with a logo or image that captures the essence of your company. The cover page has the single key contact, such as you or the CEO, and contact information, including phone number, fax number, email address, web site URL, and physical address. It should also include certain formal language such as:

This business plan has been prepared by the management of Your Company, and is being furnished to select individuals for the sole purpose of providing potential financing to the Company. This business plan is a confidential document that contains ideas, concepts, methods, and other proprietary information. Readers are to treat the information contained herein as confidential and may not copy any of these materials without the written permission of the Company.

That's just the start of the legal points you need to insert. You also need boiler plate language, probably a page or two in length, that protects your company from various securities law issues. For instance, if you say in the plan that investors will make millions, but you actually go bankrupt, an investor could then sue you for fraud and misrepresentation. This is avoided when you use certain magic words such as:

None of the projections given in this document should be taken as forecasts or promises, nor should they be taken as implying any indication, assurance, or guarantee that those assumptions are correct or exhaustive. Accordingly, neither Your Company nor any of its agents shall be liable for any direct, indirect, or consequential loss or damage suffered by any person relying on any statements in or omitted from this document, and any such liability is expressly excluded.

The bottom line here is that once your business plan is ready, you need to have it reviewed by a business attorney to make sure you aren't turning your own dream into a nightmare by opening yourself up to shareholder lawsuits, defamation actions from competitors, Securities and Exchange Commission (SEC) investigation for criminal violations of the US Securities Act, the Sarbanes-Oxley Act for protection of investors, and more.

There are full business plans and business plan assembling software that you can download or purchase, where you fill in the blanks to fit your company. In the end, however, once you understand what is needed, you will probably get the best result by writing it yourself.

Patents Expert Pointer

You can inexpensively download the necessary legal language to protect your company and insert it after the cover page. Similar downloads are available if you're looking for investors from other countries.

The Executive Summary

The first piece is the executive summary that takes one or two pages to lay out the purpose of the business, the market you plan to take on, the intellectual property that you have, the experience of your team, the scale and timing of your projected business success, the amount of money you hope to raise, and the percentage of the company you are willing to give up in exchange for the investment. If a VC reads this, he will at least know if you fit the criteria of the type of investment his firm is looking for.

VCs may be looking for specific types of companies to invest in. They may want their money going into Internet startups with two years to full success and IPO, where investors have at least 51 percent ownership (hence total control) of the business. Or they may have investors looking for a ten-year time frame with novel biotechnology with at least a 10 percent share and a seat on the board of directors for the VC firm. The executive summary will answer all of these critical questions without the investor even looking at what the actual technology is.

VCs also check that your management team meets their requirements. They may invest only in companies that have a CEO who has been involved in at least five previous startups and are located within 5 miles of the VC firm's office (so they can easily drop in to check on you to make sure your representations about business progress are verifiable).

The business plan itself can easily be over a hundred pages.

Description of the Business

Your description of your business should be about twenty-five to thirty double-spaced pages covering the following issues:

➤ Purpose of the Company: This states your main objectives. It could explain how you intend to seize the major portion of an existing market, create a new market for an entirely new type of product, become a technology licensing entity, or develop manufacturing.

➤ Existing Industry: This provides some background to readers who know nothing at all about your business. If the startup is based on a new method of doing plumbing repairs, then you need to lay out some information about the scale of the plumbing industry and how it works (part suppliers, distributors, large multistate plumbing firms, small local plumbers, etc.). You can also discuss how recent new technologies have affected this industry.

➤ New Technology and IP: Here you can describe your invention and how it fits into the history of technology in your industry. Is it a much-wanted critical improvement or is it a revolutionary change that will alter the way in which everyone does business in your field? You can also explain if you own or license your IP (intellectual property), and explain any other relevant parties.

➤ Aspects of Business: Describe exactly how your business will work. Describe the service—will you have receptionists, an in-house call center, downloadable manuals, field reps, an Amazon.com storefront, etc.? Explain the logistics and associated technologies you will rely on.

➤ Management: Here you can provide more details about each member of the management team (you can have all the CVs in the appendix). Describe what the responsibilities of each manager will be, and explain who will be full-time and who will be part-time. Will you be using any critical consultants or specialized vendors to extend the range of your core in-house team? Mention the accounting firm, legal consultants, marketing firms, and web consultants you plan to work with.

> Operations: How many people do you expect to employ; what kind of building or facility will you rent, lease, or buy? What are the real estate and location requirements? What is the availability and typical salary of the kinds of employees you will need? What steps have you taken to arrange for the various types of insurance you will need? Where will you bank and what services has the bank offered to provide?

> Regulatory Issues: Explain the various local, state, or federal regulatory or licensing issues that impact your business. Explain how you plan to handle these various issue.

> Intellectual Property Issues: Here you need to cover details about the process by which your patent was granted and its strength. You might obtain and attach an analysis by an independent patent attorney from a recognized technology law firm that states that a further evaluation shows that your patent continues to look good on novelty and definiteness so that you are likely to prevail against infringers. You can also discuss how you will use trade secrets or trademarks to further protect the company's IP, and mention initiatives you may have in place to generate new patentable technology.

> Objectives: Here you state the speed and scale of the growth you expect to achieve, and back this up with numbers and data that show that your plan is reasonable. You also need to explain why you will succeed. You can't say that there are 500 million potential customers in China and we expect to sell one device to each person for $10, generating $5 billion in sales unless you explain why all these people reasonably can be expected to buy your product. You should explain, however, why your business is expandable to large scale and whether or not it can be operated throughout the United States and internationally.

Market Analysis

Having made a first pass in the description of the business, you now need to supply more detail for each of the major elements. It may seem unfair, but the VC may hand your plan to someone who is an opinionated expert on failing vs. successful market analyses by start-ups. You have to imagine that this person exists and work to impress him so that you get a thumbs up on understanding your market.

This section needs to discuss who the customers are and whether there are wholesalers and retailers you expect to work through. You need to explain the existing structure of the broader market and how cash flows into it and through it so that you can explain how your business will fit into this framework.

Describe who you expect your key competitors will be and your plan to beat them. Tell the reader why you will take the market share and how you will handle your competitors' expected responses to your arrival in their market.

Structure of Operations

You need to explain how operations and services are usually organized for businesses of your general type:

➤ Who employs whom

➤ Their various business models

➤ Which model you've chosen and why

➤ Who actually pays for the products and services—consumers, insurance companies, government contracts

Branding and Marketing Strategy

In the branding and marketing strategy section, explain how you will establish and build your brand, and how you want it to be perceived. Describe the impact brand names have in your field of business and how customers are usually identified and accessed.

Advertising and Promotion

Here you detail the sort of funds you expect to have, where you'll advertise and whether celebrity sponsorships are appropriate. Does it make any sense to get involved in Groupon or a similar online coupon services?

Marketing Plan

Your marketing plan should describe how you will get customers to understand that you are different from your competitors, what your plan tactics and strategies for product differentiation and market segmentation are, and the different channels you can work through.

If you have a medical service, describe whether you plan to reach private practice physicians in their offices, work through physical therapy centers, provide teaching and conference donations with a resulting booth at major academic medical meetings, or buy ads with online support groups associated with various diseases.

Risks

You need to show the potential investors that you are already aware of the many risks and challenges you will face and that you have already planned or taken steps to counter or reduce those risks. Typical challenges to discuss include the following:

➤ Difficulty with scale-up in case of success

➤ Resistance to new products

➤ Patent evasion and knock-offs

➤ Quality problems with your own services

➤ Difficulty collecting payments

➤ Problems hiring and retaining quality staff

Finances

This may be the part that you know the least about but it may be the number one focus for potential investors. You are going to need to create complex spreadsheets that show that you understand cash flows and the role of capital in the operation of your business. Projections of profits and losses, and your balance sheet may seem highly conjectural, but you need to show that you have at least thought this through and have financial models that make sense in the setting of the information provided in the rest of the business plan. Fundamental budgeting for computer systems, personnel, insurance, advertising, legal advice, consulting, and IP development need to be in place. You can attach the various spreadsheets in the appendices.

Appendices

Most of the appendices are mentioned above, but in many cases investors will want to see all of the financial spreadsheets, the CVs of the management team, the patent, and an explanation of the patent. Any media reports about your technology, key managers, or your business can be attached as well as scientific publications that relate to the invention or its potential impact on the industry.

I once had a technology written up in the "Technology Quarterly" of *The Economist* and that led directly and rapidly to a large investment by a British bank's new venture capital fund. The bankers really didn't look too hard at anything else—they knew that everyone in the world in the financial industry had just read about how promising our technology was, and they wanted to show how clever they were by getting in as investors before anyone else could make a move. It did put their new fund on the map, and publicity about their investment led to a flood of additional top-grade proposals coming before them.

Conclusion

The conclusion should be a brief recap of the most important points throughout the business plan condensed into just one or two paragraphs.

Once you are done writing the business plan, read the whole thing many times. Have others read it. Update, change, and optimize as needed. Finally, have it available on CD or as a PDF that is no bigger than 10 megabytes so it can be received as an attachment by most email programs.

Be prepared to print and bind your business plan for those who like paper. One modern trick is to upload to the FedEx Office server and have it printed at a store near the VC wherever they are in the country so it can be delivered by hand on the same day within an hour or two of your call.

If a VC asks for a printed copy, don't tell him you've already sent it electronically and he can print it himself on a regular printer. If you do tell him that, then you really don't understand your place in this process.

CHAPTER 16

 # The Role of Patents in Business Strategy

> ### In This Chapter
>
> ➤ How patents affect the value of a company
>
> ➤ Planning an intellectual property portfolio
>
> ➤ Picket, design around, and blocking patents

Inventors are often focused on the value of the product or process they have made possible. For a company, however, the value of the patent is important as well. This value can be a good deal greater than the value of the invention itself.

How a Patent Enhances the Value of a Company

A brand new startup company that has nothing in it but you as the sole shareholder and an exclusive license to your university-owned patent can have considerable value. If a third party wanted to buy the company, it could have to pay a great deal more than if it were just trying to buy the patent from the university.

One reason for this is that the company immediately includes the possibility of investing in both you and your patent. If you have started a software company and have hired a programmer who is working on a second patentable invention and a business manager who has already secured two big contracts for volume sales of your invention, then the value of your company has just shot up again.

When a patent stands alone at a university, there may be no good access to your invention for any investor. An investor may have money but no ability to turn your patent into an operating business. If there is an existing business in your field that wants to buy or license your patent from the university, then the price will be low because the company will see the patent as just one more asset along with its existing strength. It will see the patent as an opportunity to add to its existing value. However, it also may know that without its action, the university has no way to make any money with the patent. These considerations can lead the university to sell or license the patent below its value.

Once you have an operating business that shows signs of actually monetizing the patent, then its value is much more accessible for a financier. If you have put together a believable business with a good business plan, the investor can reasonably expect to buy some shares, walk away, and then come back a few years later and get a larger amount back in return. The existence of the company has turned your invention into an investable commodity.

The Value of a Company vs the Value of a Patent

Another perspective on company value becomes clear when you compare a new company with an impressive patent to a company such as a new bakery. The value of the bakery depends mostly on local competitive business factors. The value of the technology startup, however, will often reflect the full potential value of the patent internationally.

If there is a market in a particular city for $100,000 in plumbing specialty product sales per year and there are already four successful plumbing connector companies, then the new company—if it is just as good as its competitors—might expect to get one-fifth of that market. Out of that $20,000 in sales, about 10 percent is for pipe connectors, so you can expect additional sales of $1,000 per year by adding a connector product.

For a new startup with a good patent on a great new type of pipe connector, the potential value may be determined by the entire value of the market throughout the country, and if you have international rights, in other countries as well. If there is a good argument that the patent will result in a monopoly position in that market, then the potential value could be very large. The whole market for the pipe connectors could be $10 million per year.

Establishing Value Before Anything Is on Sale

Let's say that your plumbing patent is so good, that there is a reasonable case to be made that everyone will have to have your product. Right from the start, you can say that you will have every plumbing supply customer in the country coming into your stores or buying from you online. If your stores will offer a full range of plumbing supplies (unpatented) but is the sole provider of the new invention, then the resulting value may include all the sales you will get from unpatented products bought by people coming into your store. Now your invention that has a potential sales value of $10 million per year, may mean that you get 30 percent of all plumbing product sales since everyone is suddenly buying from you anyhow. This suggests a business with over $30 million per year in sales.

Patent Vocab

Valuation is an assessment of the market price of your company. Once a company is publicly trading on the stock market, the public sets the valuation based on how many shares are owned and their price. However, for early stage companies, valuation has to be estimated and argued based on the apparent value of the intellectual property and the potential market size. One looks at an estimated annual profit and multiplies that by five or ten years to project how much profit a given business would generate for a new owner. This forms the basis for estimating a purchase price for the company or for determining what percentage of the company an investor should get in return for an investment.

When someone values a company for purchase or investment, he or she is likely to consider the value of the company over a ten-year period. If you can reasonably predict the $30 million per year in sales, then the valuation could be $300 million. If your success as an innovator convinces investors that you will be producing a string of important new inventions, you can make an argument that your further innovations will win you 50 percent of the entire market. If that argument is convincing, it could be reflected in the valuation of your company as well.

Buying Time to Bring a Product to Market

One reason a patent in your control today can get you value before the product actually is ready to go on sale is because the investor can understand that the twenty-year monopoly granted by the government will give your business a competitive edge as soon as it opens, and that the competitive lead will be maintained for many years.

One of the fundamental arguments for the patent system is based on the idea that the inventor will require time and considerable expense to develop his invention into an actual product. There may be additional research and development work to fine-tune the invention. Manufacturing must be established and scale-up developed. Marketing, packaging, sales force development—all of this requires up-front funding that will need to be recovered from later sales of the product.

Inventor Beware

The monopoly provided to inventors is supposed to help compensate for all the time it takes to develop and prepare the marketplace for a new invention. Potential customers have to have time to learn about how the new product can be used in their businesses.

When an infringer moves in after the fact, he is stealing your time and the expense spent in research and development, as well as your time and the expense spent in market preparation and advertising to create an interest in the new product. For these reasons, infringement is more likely to occur many years after the invention is patented and in the marketplace.

A new product has to be introduced to the marketplace. Consumers will need to learn about it. The market itself may have to reconfigure itself to accommodate the impact of your innovation; old patterns of buying and working may need to change. There may be considerable need to educate potential consumers.

If a competitor could come along five years later, after you've finally perfected the product and once you've educated the public and developed strong sales, think about what the competitor could do. He would not need to make the kind of investment that you made—your original research, the costs of writing and prosecuting the patent, the development and marketing, educating the public, developing the market. All the competitor has to do is make a knockoff of your product, and he doesn't have to recoup all the prior expenses. Therefore, he can profitably sell your product for a far lower price than you can.

It is exactly for the purpose of encouraging all this preparation and investment that the patent laws provide the twenty-year monopoly. The fact that your company owns intellectual property with a granted patent tells the investor that you have the ability to recoup the investment based on monopoly pricing of later sales of the product.

How a Company Strengthens Its Patent Position

An important focus of your technology company should be making the most out of your innovative capabilities by making additional related inventions. This is not all a matter of having more great flash of genius moments; rather it is often a matter of cold, hard, rational business strategy.

Broad vs Narrow Patents

The effect of having time to develop your invention is greatest with a broad, or foundational, patent that more or less creates a new field or carves out a significant segment of the market in your industry. Broad patents are harder to win in the patent office and are easier to attack for invalidity in an infringement litigation. However, these are the patents that can convince an investor that your business will be a major force in your industry over the next twenty years.

A narrow patent covering a solidly valuable product is helpful, but it will be easier for competitors to invent around you so that the narrow patent's impact on the market will be smaller. In the MRI patent in the appendix, the claims go to creating an entirely new capability: the ability to image nerves in the body and neural tracts in the brain (MR neurography and tractography).

This is a broad patent that lays out a variety of elements of an entirely new field. There is essentially no way to invent around it since any improvement will still be within the scope of the original patent. You can invent and patent improvements, but you would need a license under the broad patent along with your new invention.

By comparison, in the field of MRI for blood vessels (MR angiography), there are dozens and dozens of patents for particular ways of doing MR angiography. No company selling MR angiography software can really do business unless it has a license to use or own one of these patents. However, having control of one patent doesn't provide any particular power relative to the whole field. The various MR angiography patents are narrow. They are easy to get granted and easy to defend, but they offer only slight advantages and differences relative to each other in terms of the angiograms they produce, even if the actual software elements are dramatically different in the way they operate.

Having unique internal software instructions wins the day in the patent office or during an invalidity battle in federal district court. However, if the resulting product (the angiogram or representation of the human blood vessel) is just about the same as the image produced by other very different MR angiography software, then there is little market power accomplished by having the monopoly grant to the one kind of MR angiography in that particular patent.

Post and Picket Patent Strategies

You have your invention and patent, but the degree to which it is broad or narrow is built into it. There is nothing you can do after the fact to change the "footprint" of your patent in the technology landscape of your field.

As pointed out in the previous section, it makes more sense to launch a business with a broad patent, although the risks of getting the patent and defending it are greater. However, even with a narrow patent, you can build a company around the IP if you keep in mind a post and picket patent strategy.

Patents Expert Pointer

The patent expert has to think at the level of an IP portfolio. You can act to improve the power and value of your major invention by making additional inventions and patent filings that complement the original invention. A series of narrow patents can be assembled into a virtual "picket fence," either by a competitor trying to block your way forward, or by you trying to protect and empower the strength of your core IP asset.

The ideal is to have a core patent, the post, that has broad, powerful protection. Then you go about inventing additional unique, narrow variants on your original idea. You could end up with ten narrow patents that actually protect specific individual products.

You have an advantage because of your understanding of the fundamental core invention. For each product you develop, you include a unique aspect based on a narrow patent. With a number of narrow patents that form a ring of defense around your broad patent, you force an infringing competitor to confront numerous obstacles to trying to invent around your narrow patents on the one hand, while having less to gain by invalidating your core broad patent on the other hand.

In some cases, it is possible to start with a narrow patent so that you can at least build your company and begin to operate in the marketplace. You can then, hopefully, conceive of a broader invention to patent, and then go on further with other narrow picket patents.

Fighting Off Your Competitor's Fence Patents and Design-Around Patents

There are defensive patenting strategies that may be employed by others against you. One of these strategies is called a fence patent plan. This is the counterpoint to your picket patents. The competitor sees your core broad patent and starts filing narrow patents that cover every imaginable improvement on your invention. Each of these seems to offer a better product than what you have in the marketplace. You can use your broad patent to prevent the competitor from using the fence patents, but you will have your own progress blocked. The idea is to force you to license the fence patents from your competitor because of their value to you in making a better product. This gives the competitor an opportunity to try to get you to cross-license your patent to the competitor. The overall result is to cancel out the value of your monopoly right.

Another strategy you need to fight against is an effort by a competitor to use patents in his efforts to get around your patent. In the marketplace, it is common for a company to try to design around a narrow patent. If the competitor can actually get a patent for a design around, then this becomes a great obstacle for you—particularly if he gets around the core broad patent. To counter this, you need to give consideration to writing applications that beat your competitor in the race for a design around.

Blocking Patents and Publications

Keep in mind that you already have your patents granted if you are a true innovator. For your strategy to counter your competitor's fence and design-around patents, all you have to do is write out and make public the possible bases for the fence or design around. This is a blocking strategy. Yes, if you have the resources, you can file and prosecute to get these patents yourself. However, you may be able to accomplish the necessary business effect without the time and expense of getting the actual blocking patents. If you make the ideas public before the competitor files his own applications, then your publication becomes prior art (see Chapter 8). Once it is public prior art, you can't patent it, but neither can your competitor.

PART FIVE

ATTACKS, INFRINGEMENT, AND PATENT LITIGATION

T. A. EDISON.
INCANDESCING ELECTRIC LAMP.

No. 401,646.   Patented Apr. 16, 1889.

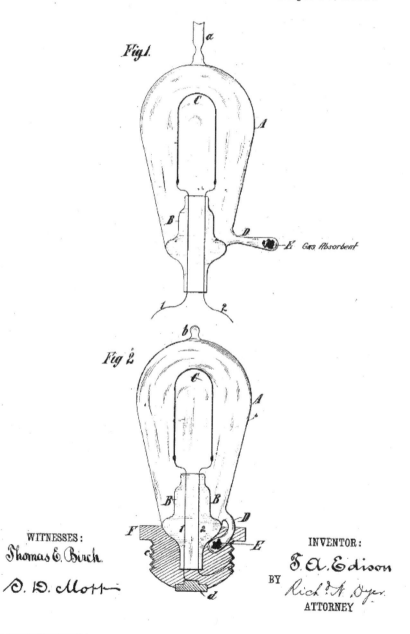

Fig.1

E' Gas Absorbent

Fig 2

 Reassessing a Granted Patent

> ## In This Chapter
>
> ➤ Addressing errors in a granted patent
>
> ➤ Strengthening a patent
>
> ➤ Attacks on a patent

In concept, you submit your completed patent application, the patent office does its examination, you then make any necessary amendments to the claims, and, if all goes well, the patent is granted and issued.

However, as the twenty-two decades have passed since Thomas Jefferson established this framework, those subject to patent infringement litigation have complained endlessly to Congress about having to fend off poor-quality patents that should not have been granted in the first place. In a series of steps, Congress has responded to these complaints.

Reexamining Granted Patents

Major changes were made in 1980 that allowed for reexamination of patents after they were issued in some situations. Additional opportunities were added in 1999 for those sued for infringement. Most recently, the AIA has greatly expanded the range and complexity of attacks on a patent that an infringer can launch at very low expense instead of actually responding to litigation to court.

The underlying idea is that litigation is expensive, but a great deal can be done by the patent office at government expense. However, as the reexamination and review process has expanded, there is more intervention by counsel for the patent holder and the opponent, so that the effect is increasingly becoming one of just turning a three-year litigation into three years of reexamination followed by three years of litigation—all paid for by the patent holder.

Although proponents of the AIA have argued that they want to promote invention, the new system imposes many additional burdens on the inventor and favors the infringer in many ways.

Why does Congress want to promote the rights of patent infringers? The catch phrase is patent quality. The idea is that the patent office is so overburdened that it can't do a sufficient job of examination. Unfortunately, the new system seems to provide a great deal more work for the patent office to do but applies deadlines to try to make it work faster.

That being said, my own experience with the reexamination process was very favorable. A party that was certainly a blatant, shameless infringer requested a reexamination of one of my patents (see the appendix) apparently as a way to slow down the patent infringement litigation in progress. The director of the patent office reviewed the submission and decided to reject the request.

This reaffirmation of the patent was helpful in quieting other potential infringers who may have wrongly convinced themselves that the patent would blow away and not have to be faced in court. It also encouraged the infringing party to settle the litigation and take a license.

Fixing Errors and Problems: Impact of the AIA

Getting a patent granted is a long, complex process taking tremendous effort over a number of years. First, you have all the years of work of inventing the invention, then the numerous hours drafting the patent application and fine-tuning it with the patent attorney. Then there are more years as it is submitted and then shepherded along through the complex patent examination process.

During examination, the examiner makes a series of objections and requires various amendments and changes. Finally if all goes well, your patent is granted. Then one day, after all your work and waiting, at long last, you get your first look at the final granted, issued patent. Right there on the top of the first page is your name in bold type as the inventor, but you notice immediately that your name is spelled incorrectly.

This is just one of a variety of errors or problems that the inventor or patent owner may discover. Depending on the timing and severity of the problem, the patent office has four different ways to respond.

Minor Corrections of a Newly Issued Patent

A misspelled name, some typos, a dropped zero in a number that makes a certain measure ten times smaller than it should be—these are all typographical errors, simple mistakes, dropped words, or printing errors. Some may have little effect other than to mar the magnificence of the document; others may significantly change the meaning.

Unlike books and journal articles in the publishing and academic worlds, there is no page proof or proofing stage when you can review your patent for errors. It gets printed and released, and only then do you get to inspect it. Usually, careful proofreading should take place immediately after the inventor and attorney first get to see the printed document.

If the problem is the fault of the patent office, then the USPTO may decide to issue a Certificate of Correction at its own expense. If the mistake is due to a problem in the documents submitted by the inventor or his attorney, then the patent owner is responsible for the expense of issuing the Certificate of Correction.

The patent is not corrected or reprinted, but when an official copy is requested from the patent office, any Certificates of Correction will be provided as the last few pages of the patent. In a patent litigation, the patent together with any Certificates of Correction make up the "true and correct" patent.

This may seem like an insufficient fix. The inventor would probably like to see the patent reprinted with the correct information; however, that is not going to happen. The issued patent is like a printed journal article or book. It is typeset, bound in volumes, and released electronically around the world. It cannot simply be recalled and reprinted with corrections. Even if patents are someday all electronic and easily correctable, the problem might be that a competitor had gotten a copy the day it was issued and adjusted his manufacturing processes to avoid infringement based on the erroneous information. The original uncorrected patent can't just disappear.

Reissue for Significant Changes

If the problem in the patent is a minor clerical or typographical error, then the Certificate of Correction provides a quick and relatively inexpensive fix. However, a variety of more serious problems could turn up.

Over the years since reissuing a patent was first made possible in 1832, the process has been fine-tuned to prevent abuses. In its current form, a patent reissue is appropriate when a significant mistake in the written specification, drawings, or claims is discovered after the patent is granted.

The effect of the mistake should be so significant that the whole patent may be inoperative or invalid without the correction. When a patent owner decides to request a reissue, he has to agree to surrender the original patent. It is cancelled, and there is no guarantee that the reissue process will result in a new corrected patent being reissued.

Inventor Beware

When a patent reissue is requested, it is possible that the examiner will notice an error he had made in his original findings, resulting in a decision not to reissue a given claim, even though it was not one of the claims involved with the reissue.

A mistake in the written text disclosing the invention or in the drawings could affect whether a particular claim should have been granted or may require that a given claim be rewritten. These are generally mistakes, and the corrections require significant further examination of the patent application by the examiner.

The result may be to have the original patent withdrawn and a new reissued patent granted with a different number beginning with the letters Re just before the number. This does not extend the patent life, but it does replace the original patent.

The rules of reissue are carefully drawn to avoid allowing the inventor to intentionally add new material to the original invention. After all, it has now been a year or two (or more) since the application was submitted. The inventor may have discovered a small improvement or adjustment that greatly improves the product or would make the patent more immune to attack by an infringer. However, that sort of change is not allowed. The problems dealt with in a reissue should always be a mistake and not an intentional change or improvement.

The two principal types of errors in patent claims are claiming too much or claiming too little based on the invention described in the text of the description.

If a claim is too broad to be properly supported by the text in the specification, then it will be easier for an infringer to attack and invalidate the claim. If the patent owner realizes this before any litigation starts, then the reissue process provides a way to get the patent reissued with narrower, more limited, and therefore stronger claims.

It is also possible to get a "broadening" reissue. In this case, the patent attorney mistakenly did not claim as much as he could have. The reissue process can be used to get a more extensive claim issued. However, any broadening reissue must be applied for within two years of the date of grant of the patent.

One problem with a broadening reissue is the question of what to do with a competitor who adjusted his manufacturing to dodge the claims in the original patent. When the broadening reissue comes out, the competitor is now infringing even though he acted with due care not to infringe the original patent as issued. Even worse, your competitor may have gotten the idea for a new product based on reading your patent description and realizing that you failed to claim something. He is making a new product based on an aspect of your invention disclosure, but you can't stop him because you didn't claim that aspect.

It turns out that the competitor generally cannot be stopped by the broadening reissue. That is considered unfair since he may have invested quite a bit of money in setting up manufacturing, marketing, and sales relying on the granted patent. This is considered a matter of intervening rights. Another manufacturer or competitor cannot start making that product after the reissue. Only something that went into production after the original patent was granted but before the reissue can win the intervening rights.

Patent Vocab

Intervening rights means that if manufacturing was started before the patent's reissue, then all the products made and sold before the reissue cannot be treated as infringing. The manufacturer is allowed to continue making this product even though it now infringes the reissue patent.

One other limitation on the broadening reissue is that you can't use this process to win back coverage that was given up during the original examination process. Suppose you started with a broad claim but had to narrow it to claim less because the examiner required it to be narrowed in order to allow it to be issued. You cannot then go back by using the reissue process to try to recapture the lost area with a re-broadened claim.

One big problem that the law describing patent reissue is supposed to deal with is the possibility that the patent owner is engaging in some sort of fraud or unfair manipulation of the patent office examination process. If a reissued patent is later used in patent litigation where the changes made in a reissue play a significant role, then the infringer will claim that there was a fraud by the inventor.

The assertion is that the inventor knew of the defect while the original examination process was taking place but intentionally failed to tell the patent examiner. The infringer claims that the only reason the reissue was requested by the inventor was that the inventor realized that the infringer would be able to invalidate the claim in court. This assertion, therefore, is that the infringer should not be forced to respond to the stronger narrowed claim or the corrected more resilient device in the new drawing. In fact, if any fraud can be proven, even if it represents only a small change, then the entire patent can be judged to be unenforceable due to fraud on the patent office, and the infringer has thereby permanently destroyed the whole patent.

Reexamination Requested by Inventor

The reissue process described in the preceding section allows the inventor to come forward with mistakes made in the patent application itself such as in the wording of the descriptive text, the drawings, or the scope (breadth) of the claims. However, reissue does not allow for the inventor to tell the patent office about any additional prior art that could affect the patentability of a given claim or even the patentability of the entire invention.

The scenario here is that after the patent is granted, the inventor learns of something in the prior art that an infringer could bring into court to invalidate the entire patent or a key claim. However, the inventor realizes that if the patent claims were rewritten, then they could dodge the newly discovered prior art so that they would be narrower but also immune to attack if an infringer should also discover that additional prior art.

Starting in 1980, the US patent law has permitted the inventor to file a request for reexamination with the patent office. The inventor has to make the case that the newly discovered prior art has a sufficient impact as to raise a substantial new question of patentability (SNQP). As detailed later in this chapter, this kind of reexamination can be requested by anyone, not just the inventor or patent owner. This old, or ex parte, type of reexamination is based on looking only at prior patents and prior written publications.

The inventor submits the request to the patent office along with the patent or printed publication and points out exactly how the additional references affect the novelty or nonobviousness of each claim involved. The patent office then has three months to review the request to decide if it really does raise a substantial new question of patentability.

If the patent office decides that additional prior art does not meet this level of concern, the request is rejected. This is actually excellent news for the inventor. If an infringer tries to raise this prior art in a patent infringement litigation, then the judge will see that the patent office has already looked at this and decided that it does not affect patentability.

If, on the other hand, the patent office decides that there is an SNQP, then a reexamination process is started. This process could also result in a determination that the original claim is not affected. However, if a change is required, then the original claim in the issued patent is no longer in force. Any infringement under the original claim can no longer be challenged in court. A new claim will be issued, and it will go into force from the date of completion of the reexamination process but with the same date of expiration as the original patent.

Supplemental Examination Requested by Inventor

The AIA adds a more-extensive option for inventors only, and this is the supplemental examination. It differs from the older reexamination process in that only inventors and patent owners can use it. It is also more extensive than reexamination because the inventor can provide not only patents and printed publications, but also evidence of prior public use or other kinds of verbal presentations and disclosures. The supplemental examination process can also be used to point out that the original examination did not look carefully enough at some prior art that was given to the examiner initially.

AIA: Change of Rule

Supplemental examination is a new process in which an inventor can ask the patent office to reassess the patent in light of prior art that was missed initially or that the inventor feels should be looked at in more detail. It is a relatively inexpensive way of heading off a challenge from an opponent in court during a full-scale litigation.

One reason an inventor would seek a supplemental examination is to try to head off a potential attack from an infringer alleging fraud. This would occur if the inventor realized that he innocently failed to provide a document to the patent office even though it can be shown that he knew of the document at the time of the original examination.

The infringer will claim that the inventor acted intentionally to hide the additional prior art and thereby defrauded the patent office. However, if there is no proof of intentional fraud, the inventor can present the document to the patent office as part of a supplemental examination request.

The patent office could decide that fraud was involved and take action. But if the patent office decides that the information was innocently not provided (by error or due to failure to realize the relevance), then no infringer can use that particular omission as the basis of a fraud assertion in the future.

For both supplemental examination and an inventor-initiated ex parte reexamination, the purpose is to use the patent office to correct, improve, and strengthen the patent. Neither the disclosure text describing the invention nor the drawings can be changed in any way, but the claims can be amended.

Once litigation is commenced, the inventor cannot use a supplemental examination to avoid facing the problem in court. However, if the inventor discovers the problem in advance, then supplemental examination and reexamination provide a comparatively safe means to deal with the problem without giving an infringer's attorney a hearing in front of a judge to attack and invalidate.

When the Original Grant is Challenged

Individuals and companies that are sued for patent infringement act like anyone else under attack. They will defend by returning fire on the attacker. In a patent infringement court room, this means employing the full array of legal processes and challenges, motions, discovery actions, and more that are described in the next chapter. The entity accused of infringing can be forced to spend millions in legal costs defending itself, and could face a judgment of many more millions. Large areas of business can be disrupted. Further, even without litigation, a granted patent may lead a business to avoid manufacturing and selling products of a particular type in order to avoid being sued for infringement. All of this can mean millions of dollars of risk.

Infringers usually win litigations. In the majority of patent litigations, the district court judge or the jury finds some or all of the claims to be invalid. The infringer has won, but that doesn't recoup the cost of the litigation or the cost of any lost business opportunity the infringer suffered in trying to avoid infringement. Faced with repeated episodes of losing millions of dollars in order to win against an invalid patent, many companies have gone to Congress to complain that the patent office is too inventor friendly and overworked, and that it is granting many poor-quality patents that should never have been granted in the first place.

In truth, the high intensity of a patent litigation applies large amounts of intensive legal effort over years in a way that goes far beyond what the patent office can muster. It is not clear that the United States can afford to raise the intensity and quality of patent examination by the USPTO to the level of scrutiny applied in full-scale litigation. However, Congress has made a series of changes in the patent laws over the years so that the patent office can do more to try to help potential infringers avoid these enormous costs of patent litigation. The AIA has further expanded the options considerably. There are now four major routes a company can use to try to shut down a competitor's patent without having to face the full expense of a court action.

How the AIA Changes the Game

The AIA greatly expanded the opportunity for a competitor to raise concerns during the initial examination before a patent even issues; this is the pre-issuance submission. Next, it provided a new opportunity for a competitor to attack the patent within the first nine months after it is granted by using a wide variety of arguments and lines of evidence; this is the post-grant review. This new process even allows the competitor to propose entirely new legal theories that could be used to render one or more of the claims as unpatentable. The AIA also changed the standards for a type of attack formerly called an inter partes reexamination, which is now designated an inter partes review.

AIA: Change of Rule

Pre-issuance submission, post-issuance submission, and post-grant review are new opportunities for additional evidence to be raised and early attacks to be more easily launched close to the time of issuance of a patent. The inventor can use these to help refine and strengthen the patent, but more likely these are methods for an opponent to try to nip the patent in the bud without the full-scale expense of litigation and before major business changes are made in respect of the new patent.

The AIA established a new Patent Trial and Appeal Board (PTAB) that replaces the older Board of Patent Appeals and Interferences (BPAI). Examinations and reexaminations are carried out by experienced patent examiners in the Central Reexamination Unit (CRU), while reviews are carried out by administrative law judges in the PTAB.

The new examination and reexamination processes are similar to the basic patent examination process. The review proceedings are now closer to the format of litigation— carried out in front of judges rather than examiners, allowing for legal discovery and depositions. The new submissions option simply allows an opponent to attach relevant new information to the patent document with limited examination of the new information.

Pre-Issuance Submissions and Protests

The new pre-issuance submission process allows for a competitor to try to stop a patent from being granted in the first place. In the past, there was a little-used process called a protest that allowed a competitor to take a shot at the application. The old protest was only useful if the inventor released information about the application while it was still being held confidential by the patent office.

If a competitor learned of the application and also knew the application number, then the competitor could submit any information of any type (documents, transcripts of oral proceedings, reports of public use of the invention, etc.) to the examiner. The identity of the person or party filing the protest could be kept secret, and no fee was required (unless the same source filed multiple protests).

Once the patent office published the patent application, anyone wanting to file a protest had to actually write to the inventor and get permission.

Under the AIA, there is now an opportunity for a pre-issuance submission by any member of the public. This pre-issuance submission is a bit more limited than a protest in that only printed publications, published patent applications, and patents can be included, and there is always a small fee. However, it can be filed at any time within six months after the date that the patent office publishes the application unless the patent is actually granted before the six months elapses.

To file a protest, the competitor had to learn of the application while it was confidential, but the new process occurs after the patent office makes the application public but before it is granted. Fortunately for the inventor, the patent examiner is likely to inform the inventor of the new material and get his response if it results in any new objections to granting any claims.

Patent Vocab

A file wrapper is a folder at the USPTO containing all the documents and correspondences related to a particular patent application.

Post-Issuance Submissions by Opponents

Once the patent is granted, there is an opportunity for any party, including the inventor, to formally notify the patent office of additional prior art in the form of printed publications, published patent applications, and patents. This information can be assembled into the file wrapper of the patent so that it is immediately available in the case of any future reexamination or review or in case of any future litigation.

An inventor can use this process to help avoid later charges that he intentionally kept some prior art secret. By using the post-issuance submission process, the inventor can show good faith by making the material available, but can avoid triggering a reexamination.

A competitor can use the process to inexpensively and effectively weaken or destroy the patent without taking the time or expending any effort to actually have the information fully examined or reviewed. The patent office will look at information provided in the post-issuance submission to determine if the information is relevant so that it should be included, but that is all the analysis that is done. This would be useful if the competitor has discovered prior art that seems likely to immediately destroy the patent if any action is taken.

Post-Grant Review

The most dramatic change in the new patent rules for attacks on a patent without actual litigation is the post-grant review. This is something that an opponent must launch within the first nine months after the patent is granted. However, it is wide ranging because it allows the opponent to use an array of methods of attack. The other types of reexamination and review are limited to asking the USPTO examiners or the PTAB to look at published documents, published patent applications, and granted patents. A request for post-grant review, however, can also include statements (affidavits) reporting on prior public use or even transcripts of oral presentations as well as transcripts of other court proceedings. It can challenge enablement, arguing that the description is not sufficient to make the invention work or that it is described well but just doesn't actually work.

In addition, an opponent can also propose a new legal theory that could be argued to render the new invention or one of its claims unpatentable. An example of this would be something along the lines of the effort to limit or forbid the issuance of patents on genes.

The director of the patent office has three months to review the petition for a post-grant review. The petition has to meet a fairly high standard—it has to convince the director that with the information provided, it is "more likely than not that at least one of the claims challenged in the petition is unpatentable." This is more demanding than the old SNQP standard of substantial new question of patentability used to decide whether to allow an ex parte reexamination under the old system. However, the post-grant review process also differs from older processes in that the patent office notifies the inventor of the petition and allows him to present an argument to the director as to why the petition for post-grant review should be rejected.

If the post-grant review goes forward, it is handled by the judges in the PTAB, and they are supposed to have just one year to complete the review and make a decision. If either party disagrees with the decision of the PTAB, then the matter can be appealed directly to the Court of Appeals for the Federal Circuit.

This process is not anonymous. Further, if the patent and its claims survive the challenge, then the result is gold-plating of the claims and patent. A district court judge later faced with an action for invalidation against the patent will be influenced to have much greater respect for the patent and claims. The actual party who requested the post-grant review is actually barred from raising this issue in any future efforts in court to invalidate the patent.

Ex Partes Reexamination: Getting a Redo

Since 1980, the patent law has allowed any person to request that a patent be reexamined based on prior art not considered or alleged to have been insufficiently considered in the original examination leading to grant of the patent. The petition to the director has to provide a printed publication, published patent application, or granted patent, and it has to specifically show how this prior art raises a "substantial new question of patentability."

Patent Vocab

Ex partes and inter partes refer to the level of participation your opponent chooses to take when a challenge to your patent is launched through the patent office. In an ex partes process, the opponent notifies the patent office, provides the evidence, and points out the way the evidence could affect the patent—all in an initial filing—and is then generally locked out of the process of evaluating the effect on the patent. In an inter partes process, the opponent is involved in the back and forth exchanges as a decision is reached about the impact of the challenge.

If the Director of the USPTO decides to allow the reexamination to go forward, the inventor is notified and allowed to file a response or provide amended versions of the claims to try to dodge the new information. This response is shown to the party who requested the reexamination, and that party gets to respond back.

However, from this point forward, the petitioner has no further input. This is really where the notion of ex parte comes about. The term mostly refers to the fact that after the initial round, the new information is dealt with back and forth between the examiner and the inventor. The petitioner is locked out of the discussions and can't add any further input or argument. The process can be long and slow and has dragged on for as long as three years.

This is another process wherein success by the inventor in defending against the attack results in a gold-plated patent that is going to appear stronger and better in the eyes of a judge in any future litigation. However, the party filing the petition is allowed to take another shot with the same information in a courtroom, where the full array of discovery, deposition, and argument are available.

Inter Partes Review

In 1999, US patent law was changed to allow for the petitioner to take a more active role in the reexamination process. This is the inter partes reexamination that offered an upside to the attacker, who could file arguments and responses throughout the reexamination process. The downside for the attacker was that a loss in the inter partes reexamination blocked that party from having another shot at the same issue in the courtroom (unlike ex partes reexamination, where no such bar applies).

The AIA has redesigned the inter partes reexamination and replaced it with a process called inter partes review. This now gets heard by the PTAB judges rather than by the examiners. Like the post-grant review process described above, the inter partes review gets the rapid treatment of a one-year time limit for the PTAB to complete the review. Also like the post-grant review, this new kind of review allows for some discovery and deposition evidence to be presented to the PTAB.

AIA: Change of Rule

Under the AIA, the word reexamination signals a reassessment by patent examiners, but the word review signals a reassessment in front of a judge at the new Patent Trial and Appeal Board (PTAB). The new review process under the AIA is more difficult for an opponent to set in motion than for similar proceedings in the past, but once it starts, the PTAB review is more like a trial because it allows for discovery and depositions.

The standard that has to be met in order to be allowed to proceed has been raised significantly as well. It is no longer just a "substantial new question of patentability" (requirement for an ex parte reexamination), and it is not sufficient to meet the "more likely than not" standard (requirement of the post-grant review). Rather, the petitioner must convince the director that the new information provides "a reasonable likelihood that the petitioner would prevail."

The inter partes review does resemble the ex partes reexamination in that only prior art in the form of printed publications, published patent applications, and granted patents can be considered. This does not have the wide-ranging array of allowed types of challenges that can be brought in the post-grant review.

As with the post-grant review, however, if the petitioner loses with the result that the patent and its claims survive unchanged, then the patent gets the full gold plating of additional respect by judges in future actions and the platinum plating of a bar to the petitioner against even raising this issue in any future court action (the bar is called an estoppel in legal terms). This platinum bar is useful because the petitioner is also blocked from attacking in the future with any information or argument that could have been raised in the inter partes review even if the petitioner did not actually do so.

Stays, Appeals, and Declaratory Judgments

Increasingly, with new provisions of the AIA, attacks on patents through the patent office and through the PTAB provide a full-scale alternative or parallel process by comparison with the federal court system.

Choosing Between the Courtroom and the Patent Office

Overall, costs are lower if you don't use the courts, but both sides give up the full array of legal capabilities offered by the court system. However, the disputes from both the federal court system and from the USPTO and PTAB systems eventually go to the same court if appeals are made. Everything ends up in front of the same set of twelve judges who make up the Court of Appeals of the Federal Circuit.

If a patentee or an accused infringer tries to use both systems at the same time, then the USPTO and PTAB processes usually take precedence and any court proceedings are usually put on hold (technically, a stay is granted). Then, once the issue raised in the reexamination or review is finally settled, the litigation in district federal court resumes. Of course, if the entire patent or the relevant claims have been declared unpatentable due to the USPTO and PTAB actions, then the court action is dismissed as well.

One additional twist thrown into the mix by the AIA concerns a legal action called a declaratory judgment (DJ) (discussed in more detail in the next chapter). This is an action generally feared by inventors in which a competitor makes a first strike against the inventor by forcing the patent into court when the inventor may not have wished to engage in legal action.

The inventor may just have sent a letter to a competitor informing him of the patent and offering a license. This can allow the competitor to file the DJ in his home state even if the solo inventor resides in another state and can't afford litigation in the first place. When a DJ is filed, the competitor is barred from using most of the reexamination and review processes available with the USPTO or PTAB.

In general, the USPTO and PTAB routes will be less expensive and it may be faster. However, it is just as likely that both a USPTO and a court proceeding will play out so the total time and expense just piles up, one on top of the other. The USPTO offers greater expertise with technology and patents, but the courts offer a more complete array of argument opportunities and legal maneuvers.

Impact of Appeal Rules

The various review and reexamination processes can all lead to the same appeal forum in front of the Federal Circuit, where any appeals of patent cases from the federal courts will also go. This does provide finality to critical questions that may lie at the heart of a patent dispute.

Even though the same ultimate forum is reached, however, the information put in front of the Federal Circuit judges is different with the two different routes. In concept, the information arising in the review and reexamination processes are more precise and targeted. However, information deriving from the federal district court will be more extensively argued and detailed. A Court of Appeals generally just looks through information already provided and discussed in the lower forum but does not hear or receive any new information itself.

Overview of the Risks

In the grand scheme of things, inventors will probably do better by going through the courts and accused infringers will probably prefer the review and reexamination processes. Keep in mind, however, that out of 250,000 patents granted each year, only about 3,000 end up in litigation in a given year. In the last year of the old system (2010–2011), there were about 800 ex parte reexaminations per year and about 350 inter partes reexaminations. Most of those involved in reexamination and review are the same patents being litigated.

These numbers don't mean that litigation and reexamination and review are unimportant. Rather, these processes are relevant to the most valuable and controversial patents. When you consider that a typical patent litigation can cost about $3 million in legal expenses for the inventor, you can readily see that only patents with significant value are going to be litigated. You can only win back the damages you suffer, so most patent conflicts are avoided because the amount that the inventor can win is less than the amount it will cost to prove the rights.

Supplemental reexamination and ex parte reexamination filed by the patent owner do continue to be less expensive ways for the inventor to strengthen a patent in order to pose a stronger position in licensing negotiations. For the vast majority of patents, however, the original examination and grant by the patent office are the sole and all-important tests of an invention.

1,258,730.

Patented Mar. 12, 1918.

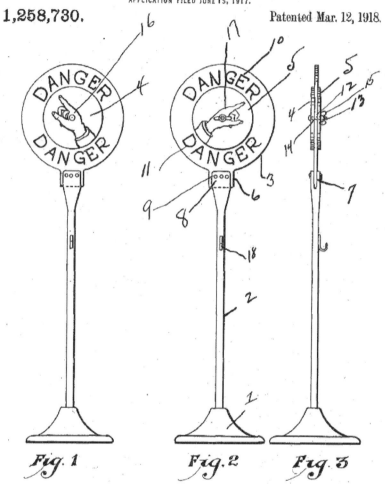

Fig. 1   Fig. 2   Fig. 3

WITNESS

R. F. Dilworth

INVENTOR.

Howard L. Wood

By Max H. Groling
his attorney

CHAPTER 18

Responding to Infringers

In This Chapter

➤ Expense of patent infringement litigation

➤ Risks of a declaratory judgment action

➤ How nonpracticing entities affect patents

➤ What can be gained from infringement litigation

A patent protects the right to make, use, sell, offer to sell, or import into the United States the invention described in the claims of the patent. If a company is making a product you invented without your permission, it may be infringing.

Recall that a US patent only covers activities in the United States, so if someone is making your invented product in Mexico or China, there is nothing you can do about it. However, if the product is then imported and sold in the United States, then the company that is importing and selling it is infringing. The customers who are buying and using the product or method are infringing the patent. A company that is advertising the covered product is infringing as well. In practice, however, there are only a few limited situations in which trying to stop someone from infringing makes good sense.

Recognizing Infringement and Considering the Impact

The first step in protecting your intellectual property is to be able to find out if someone may be infringing. In the case of some inventions, the product is so specialized that you have only one competitor who could conceivably be trying to copy it. You keep a close eye on your competitor, and you will know immediately if something is up. In the opposite situation, your patent covers something that hundreds of manufacturers and tens of thousands of mid-level users could be starting to quietly copy and use.

Maintaining Surveillance for Infringement

In many cases, a competitor engaging in patent infringement will take some steps to keep its activities quiet. If he knows of your patent, he may hope to avoid drawing attention to his activities.

In other situations, your invention covers only a small part of a complex machine, software program, or electronic device. In this case, it may be difficult to know if your invention is being used. You may even purchase and use a complex product without knowing that it contains a component that infringes your own patent. Consider the thousands of hardware and software patents involved in the components of an iPhone, for example.

Whatever your own particular situation, you must maintain some sort of surveillance to become aware of infringement. Some firms actually hire companies to maintain surveillance. With the Internet, you may be able to identify areas of suspicion and then drill down in manuals and other documents to uncover suspicious components. One reason this is done is that if a company can show that it openly and obviously infringed your patent, that you knew about it, and that you did nothing about it for years, then that delay, itself, may become a defense that can be used against you if you do finally decide to sue.

In many cases, a company having an exclusive license to make, use, and sell a patented product also has an obligation to notify the patent owner of any infringement it detects or should have detected. The owner of the patent may include this as a condition in the license.

Is Infringement Hurting or Helping You?

There are a few situations in which patent infringement can actually be beneficial to a patent owner. One situation would be an individual inventor or small company that repeatedly failed to convince various investors, manufacturers, and potential customers that a novel product would attract any sales. One day, you walk into a store and you are shocked and stunned when you see that a huge multinational corporation is manufacturing your product, and hundreds of people are lining up to buy it. In this case, the infringing corporation has created a market for your invention that you failed to create through your own efforts. You then discover that it has launched a multimillion-dollar television marketing campaign that is building a frenzy of interest in your invention.

In this sort of situation (and in other less-dramatic versions of it), the infringing company may have failed to discover your patent and therefore it may have no idea that it is infringing. You may learn that it has actually applied for and been granted patents that seem to overlap your own invention, but that the patent office somehow failed to connect to your patent during the examination process. This sort of problem is getting rarer in the age of Internet search engines, but it still occurs. This is often because the actual words used to describe a new invention may differ so that a nonexpert does not realize that the two different patents actually describe the same thing.

In some cases, experienced inventors and companies have actually taken advantage of this phenomenon by working to obtain and control what have been called submarine patents. These are patents that can be kept out of the limelight until a market grows. Then when the market in the product takes off, the owner of the submarine patent surfaces and commences litigation.

Jerome Lemelson was a prolific inventor who was granted over 650 patents in his lifetime. He appears to have perfected the art of using a variety of motions and actions to keep some of his patents in the patent office, unpublished and under examination for decades until huge markets developed. He eventually applied infringement and other actions that led him to collect over $1.3 billion in royalties before his death in 1997.

Some view him as a genius. Some view him as a villain who made few real contributions to technology or industry, but rather cruised through the intellectual property system, taking cruel advantage of its peculiarities to take money from those who actually made the real contributions. This is overly harsh, but it is true that a person who files several patents per week in dozens of fields of manufacture has little time to actually develop any of the products.

Lemelson's most famous escapade was the patent for the use of bar codes, which emerged after bar codes were in use by thousands of corporations in millions of products. The litigation that eventually wrecked the bar code patent led to reforms of the patent system that have made submarine patents effectively impossible to accomplish in the modern system.

Lemelson was able to file a series of continuation patent applications that derived priority from the initial filing in 1954, but took advantage of the fact that life of a patent at that time was measured from the date of grant. The continuations resulted in deferring the actual granting, so that he was still filing continuations thirty years later.

Under current law, the life of a patent is measured from the date of the first original filing, not from the date of grant.

More often the story is very different. An inventor has poured all of his energies and funds into developing and marketing his product. He has mortgaged his house, borrowed from friends and family, thereby going heavily into debt. His company is just starting to earn real profits and the product is now increasingly attracting widespread attention.

A large competitor learns of the product, decides to ignore the patent, and brazenly launches a direct copy at one-half the price that the inventor is charging—a price that loses money with each sale but is calculated to crush the inventor's business into bankruptcy so that he will not be able to afford to sue for patent infringement. It's a harsh world out there and there may be nothing the inventor can do but file for bankruptcy, move in with relatives and spend the rest of his life fuming about what-ifs and what-might-have-beens.

Cost of Infringement Litigation vs Cost of Infringement

Carrying a patent infringement case through to completion may generate legal expenses for the inventor of over $3 million on average. Some cost much more. A simple, straightforward patent litigation may be completed in three years, but if the value is high, there may be various appeals and remands, diversions into the patent office for reexaminations and reviews, stays and other delays that lead to ten or fifteen years of litigation or more. This can drive the cost of litigation into the $10 million range. Even in the case of total victory, it may not be possible to recover all the costs of litigation. You may win the case but be worse off financially than if you never sued.

Patents Expert Pointer

Patent infringement litigation is very expensive and the chances of winning are about 1 in 4. For these reasons, the patent expert will focus on nonconfrontational means of gaining value from a patent. Business practices such as establishing a trusted and recognized brand name, earning a reputation for high-quality service or producing a string of innovations—all provide alternative ways to publicize a patent while avoiding litigation.

If you use the benchmark of the typical $3 million litigation expense, you can get a rough idea of whether patent infringement litigation makes sense in your own case.

First, you have to honestly figure out the scale of lost sales over the past six years that were directly due to the patent infringement. For instance, in this lost profits model, you would have to show that your product sold for $110 and cost just $10 to make so that each lost sale cost you $100 in lost profits.

Then you would have to show that there was convincing evidence that 30,000 of these were sold at this price in the past six years by your infringing competitor.

Finally, you would have to show that your own business was capable of making and selling 30,000 of these and that you would have gotten all these sales if not for the infringing actions of your competitor.

If all this were true, the infringer would be liable for $3 million in damages. If you won, you could pay your attorneys the $3 million you owed them, and that would be it. You would not recover all the vast time and energy spent on the litigation and you would not actually be any better off financially—you are still out of pocket for the $3 million.

As part of the typical judgment in a patent litigation, you would be forced to give a license to your competitor allowing him to continue producing the product as long as he paid you royalties.

The weird thing about this is that if you are right and you really would have had the profits, then all the patent litigation can do is to put you in the same situation you would have been in if there had been no infringement minus the cost of the litigation. If the lost profit was $6 million, then you get $3 million, your attorneys get $3 million, and the infringer loses his ill-gotten gains. At least this has the feeling of justice.

Since the average award in a successful patent litigation is around $3 million or $4 million, there are plenty of inventors who litigated successfully who got little more out of it than the feeling of justice.

The other way that patent damages are assessed is in terms of reasonable royalties. This is more suitable to an inventor or small company that cannot make a believable case that it was capable of accomplishing all the sales itself.

The situation may be that your invention is a part used in an automobile. Since you don't make entire cars, you would be unable to sell your invention. Perhaps you have a tiny company, and no one would believe that you could set up the network of a thousand sales and service centers that your infringer controls. In these and many other situations, it is more appropriate to use the reasonable royalty approach.

In this approach, the court asks what royalty rate the infringer would likely have agreed to in a negotiation undertaken at the start of the period of infringement. If the royalty rate would have been 5 percent of the gross margin (collections after costs of manufacture), then one can see what happens.

If the infringer sold 30,000 items with a $100 gross margin, you would get $5.00 for each of these and the damages award would be 5 percent of $3 million—just $150,000. That sounds like a nice sum of money, but it doesn't amount to much if you spent $3 million in litigation costs to get the $150,000. However, in another scenario where a major multinational corporation has $10 billion in sales, the 5 percent royalty could be worth $500 million.

As you can see, the decision as to whether litigation would make sense even if you win is a matter of mathematics. You also have to factor in the consideration that the infringer wins in about 75 percent of cases. This happens because the full force of litigation in the federal courts with many millions of dollars at stake, combined with a general tendency of judges to be suspicious of plaintiffs seeking damage awards pose a powerful challenge. This may mean that a patent claim that stood up very nicely when the patent office did the exam (and granted your patent along with 250,000 others granted that year), may not stand up to the intense scrutiny of litigation.

AIA: Change of Rule

New rules limiting joinder in patent infringement cases have dealt a harsh blow to the individual inventor. In the past if there were multiple parties infringing your patent, you could usually file a single lawsuit listing all the different infringers. Under the AIA you need to file a separate lawsuit against each infringer, usually in the infringer's home jurisdiction.

The intention of Congress was to increase costs for nonpracticing entities (see page 206), but it is the individual inventor who is seriously affected by the new rules. The costs of each lawsuit are similar to the single lawsuit, but if you can afford to file only one action, you will spend three or more years on that one case and have potential damages from only one infringer, while all others are free to go on infringing.

Direct and Indirect Infringement

In some cases, the decision whether to litigate is also influenced by the potential ability of a given infringer to pay the judgment that you would win. If you were to decide to sue mid-level users for infringement, you might find that none of them has the resources to pay enough in damages to cover the cost of litigation. Imagine spending $4 million to win a $10 million judgment against a business that is nearly bankrupt from the litigation. You win, but you get nothing.

In general, each litigation against each infringer needs to be filed and litigated separately under new rules that arose as part of the AIA. Multiple similar but separate litigations might each cost considerably less than $3 million (imagine five litigations each costing just $1 million because of duplication of efforts from case to case). Nonetheless, with smaller defendants, each litigation could still be a loss in terms of collection vs. costs in case of a win.

One consequence of this logic and arithmetic is that inventors may be tempted to look for indirect infringement carried out by a large entity with an ability to pay.

In one typical pattern of indirect infringement, a large manufacturer gives its customers tools that allow them to infringe the patent and is advertising that if they use these tools, they can accomplish the function covered in your patent. This is said to be a possible case of inducing infringement. The manufacturer, itself, is not infringing. However, it may be held liable for indirect infringement.

This finding does require that you prove that the manufacturer was well aware of your patent and that it specifically intended to cause the customers to engage in infringement. This is difficult to prove.

Another type of indirect infringement is contributory infringement. In this situation, a manufacturer makes a part that doesn't have a non-infringing use but does not infringe when considered in isolation. That part is sold to a customer, who then assembles various ordinary parts along with the specially made component to make an infringing device.

The customer, who actually infringes, would be a low-value defendant with very little damages taken by himself alone. The manufacturer is not infringing any claim of the patent. However, if the manufacturer is well aware that the special part is solely for the purpose of promoting infringement, then the manufacturer could be liable for the total value of all resulting infringement by the customers.

We can consider the situation of the patent in the appendix. Here, a small number of large multinational corporations (GE, Siemens, Toshiba, Hitachi, Philips) make MRI scanners that have multiple uses in medicine. The invention in the patent described an important advance in MRI scanning that caused each of these manufacturers to redesign their hardware and software and then led them to intensively market the new capability to show off the ability of their own scanners relative to those of competitors.

Ten thousand hospitals and image centers bought the scanners for about $1 million each, or they purchased upgrades for the new technology that cost about $300,000 each. Then each hospital used the scanners to perform thousands of scans at about $1,000 each. The actual steps of the invention were carried out by individual technologists and physicians using the MRI machines and various related computers and software.

Let's suppose you were the patent holder and you filed a patent infringement suit against a single very large medical center, expending $3 million toward winning the case.

That hospital did three thousand scans with the scanner at $1,000 each, so it made about $3 million. The scans and analysis steps were performed in part by about twenty-five different physicians and technologists.

Your company owned an identical scanner across the street from the hospital, but you didn't have the brand name appeal that the medical center had, and you didn't have the "captive" referrals from the medical center's doctors. Still, if the patent had been respected, the hospital would have sent all of these patients in need of scans across the street to your facility at no special inconvenience for most of the patients.

Even with a win, it might turn out that the medical center could not pay more than one-third of the judgment. You could then repeat this process against hundreds of medical centers, and in each case recover less than the litigation costs. If you filed against the physicians, the recovery from each litigation would be even smaller.

However, if the target of the litigation was a large manufacturer, and if the manufacturer could be held liable for inducement of infringement, then the damages would be extremely large. If each customer caused $3 million of damages, and if a given manufacturer had sold 5,000 scanners, then the damages could be $15 billion—far more than the value to the manufacturer of all the scanners it sold.

The Risks of Warning an Infringer

In an ideal situation, various businesses interested in your technology would contact you and offer to take a license under your patent for an appropriate royalty. However, there are a number of reasons that may not happen.

A company may have learned of the technology and decided to develop a product without knowing that there is a patent. Even if it is aware of the patent, the company may have an opinion that it can defeat any claim against it. It may feel that your patent is weak and likely to be invalidated. In many cases the company hired an attorney to give them an opinion that your patent is likely to be invalidated in court. The company may be confident that it can overwhelm you in court with year after year of motions and reexamination requests even though the patent looks like it is strong and valid.

One of the odd tricks of the human mind is to immediately assume that a new idea or technology must have been obvious forever. Many uninventive people don't believe in inventions, assuming that technology somehow progresses steadily without any real discrete and definite acts of insight by individual inventors.

Whatever reason that the competitor gives for infringing, it will likely become necessary to take the initiative to inform the company that you believe it is infringing and invite it to take a license. The company may see this as an empty threat and a shakedown—an attempt to take money from it and take advantage of its own business success.

If all goes well, the company will offer to commence negotiations. If negotiations start, the company will likely have a dual purpose of trying to find a reasonable price for the technology and trying to learn how much of a threat you pose. You may get an offer for a license, but at a low royalty rate you're not willing to accept. If the company takes a negative view, it may dig in its heels, refuse to negotiate, and start planning a preemptive counterstrike against you.

Declaratory Judgment (DJ)

The form in which a counterstrike is carried out is an action for declaratory judgment (DJ). This is a court filing that asks the court to take a decision against your patent. There are three typical bases for such an attack: 1. invalidity, 2. unenforceability, and 3. noninfringement:

1. An assertion that the claims of your patent are invalid

2. An assertion that your patent is unenforceable due to misconduct on your part

3. An assertion that the competitor's product does not actually infringe your patent

Inventor Beware

By simply approaching a company to propose discussions about licensing your patent, you may expose yourself to a DJ lawsuit. Typically, if the company makes a low offer and you refuse, then that is sufficient to establish a controversy so that the court will allow the DJ to proceed. If you have shown elsewhere that you are prepared to litigate against infringers, then the simple fact of a written proposal to have discussions may qualify as a controversy sufficient to allow a DJ.

One motive for filing for a DJ is that the potential infringer may get to choose the location of the legal action. If you traveled across the country to go to your opponent's headquarters for a negotiation, he may be able to assert that you have sufficient contact with his city and state to force you to appear there to answer the DJ action.

Courts don't have to agree to allow a DJ action to go forward. There has to be an actual controversy. This could mean that you have told your competitor that he is infringing and have sent a demand letter telling him to cease and desist from infringing. However, even if you were pleasant and polite in offering to discuss a license, if he offers a low royalty and you disagree, he may be able to assert there is a controversy.

If you have already filed patent infringement suits against other entities and now send a letter offering a license, the opponent may be able to convince a judge that this is a veiled threat of sufficient magnitude to serve as the basis for proceeding with a DJ action.

If a DJ action does go forward, you will probably be forced to file the patent infringement lawsuit as a compulsory counterclaim. This means that if you decide to just work your way through to try to win the DJ action without filing for the full-scale patent infringement litigation, then you may be barred from changing your mind later. If you don't file for patent infringement as a counterclaim, you may end up with no way of winning any damages even if you defeat the DJ challenge.

A Patent's Worth without the Threat of Litigation

The main reason an opponent will not want to file a DJ action is the hope that litigation can be avoided altogether. This puts the inventor in a difficult situation. If you assure the competitor that there is no way in the world that you would litigate, then he has no motive to offer you anything since the patent is worthless if you are unwilling to sue on it. If you express a likelihood that an infringement action will be filed against the opponent, then he may feel that negotiations are pointless and he might as well file the DJ to help get the location (venue) of the litigation close to his place of business.

Occasionally, an infringer will threaten to sue you in state court in response to a warning on patent infringement. The argument is that you are interfering with his business—this is the basis of a tort legal action. This is just an attempt to trick you. If you have good reason to believe your patent is valid and that the opponent is infringing, then the right to protect your federally granted patent rights overrides and preempts any counteraction on a state level. The opponent's only choice if he wants to litigate in response is a DJ filing in federal court.

Patent Litigation Companies

One solution to the problem of managing the expense of patent litigation is to bring in a partner or investor who participates financially in the litigation. For instance, you may find an investor willing to pay half the costs of litigation in exchange for half of the proceeds in case of a win. This is a specialized area of investing because any potential partner has to be able to make reasonable decisions about your patent's ability to hold up against an invalidity challenge and the potential damages that can be won.

Nonpracticing Entities (NPEs)

An important financial phenomenon that has transformed the patent infringement landscape is the emergence of financially powerful nonpracticing entities (NPEs) who buy or invest in various patents purely for purposes of litigation. The reason this has so much impact is that in patent litigation, powerful and financially strong companies are likely to be able to crush an independent inventor or small company that tries to litigate a patent infringement case. When the small companies with novel technology or the independent inventors are aligned with an NPE, then the balance of power is reversed. A large company is faced with an opponent that is not only financially prepared to litigate fully, but is also extremely experienced in patent litigation.

Patent Vocab

A nonpracticing entity (NPE), also known as a patent troll, is a company that specializes in suing infringers even though it is not actually a manufacturing, producing, or inventing business. An NPE amasses litigation funds from investors, and then makes alliances with patent owners and inventors to effectively sue on their behalf.

The combination of sufficient funding and expertise make a demand letter from an NPE into something that no company, no matter how large, can afford to take lightly. NPEs often file numerous litigations against various competitors in a field in hopes that at least one of the competitors will settle and license. This puts tremendous pressure on the other competitors since the first to license is now out from under the cloud and racing to market ahead of them. If the NPE threatens to sue the customers of the entities that have not licensed, then there is a fear that the customers will chose the licensed competitor in order to avoid a risk of litigation.

A number of provisions in the AIA are intended to make things more difficult for the NPEs, but there is nothing in the AIA that really impacts them seriously. One example of such an aspect of the AIA is the rule forbidding a case filing in which numerous different competitors are named in the same lawsuit.

This is not all that effective because the additional expense of multiple separate filings is overwhelming for a solo inventor or small company acting on its own against many infringers. The NPEs are generally so well funded that the additional expense of multiple litigations has limited impact—just the cost of doing business.

Pitching to a Patent Litigation Firm

If you are considering a litigation for patent infringement, you may seek an opportunity to make a presentation to an NPE in hopes of attracting its participation. A successful presentation requires you to fully understand the strengths of your patent as they affect all stages of the proposed litigation.

You need to explain the state of the prior art that is likely to be asserted against you. The NPE will be interested in strengths against actions for invalidity, unenforceablity, and noninfringement. It will want to see a convincing estimate of the scale of damages likely to be won if the case is successful. It will also be interested in the ownership or licensing status or your own patent rights to make sure no one will be able to get in front of it in getting proceeds in case of a win. It will be important to the NPE if you are providing the legal team, and the NPE will also be interested in talking with the inventor to try to judge how good a witness the inventor will be when put in front of a jury.

The NPE will want to see a patent litigation team that appears to be able to litigate successfully. Along with this, it will be interested in the financial arrangement under which the case is proceeding. If the law firm will have a contingency right to get a significant percentage of proceeds, then the NPE will feel that the patent attorneys have a special financial interest in winning a large settlement or judgment.

A straight hourly rate arrangement may concern the NPE since it will worry that the law firm is only interested in generating large legal bills and settling for just enough to pay its own fees without seeking sufficient funds to make the case worthwhile for investors.

One Man's Troll Is Another Man's White Knight

It doesn't take much imagination to see that NPEs will stir very different emotions in patent owners as opposed to the emotions they stir up in accused infringers. Companies that have felt as if they were forced to pay large settlements due to the threat and power of NPEs have come to call NPEs patent trolls. This refers to trolls of legend who brutally demanded payment from anyone wanting to cross a particular bridge.

For the poorly funded patent holder, however, an NPE will seem like a wondrous white knight acting in the name of justice. The NPE will step in and force a huge competitor to take a license or pay a judgment in a patent infringement litigation with little expense or risk to the patent owner.

Injunctions and Damages

There are two major end points that a patent holder can seek in a patent litigation:

1. Obtain money damages in compensation for lost profits or lost royalties

2. Stop the competitor from infringing so the inventor's business can prosper under the patent monopoly

The way that an infringer is stopped is by convincing the court to issue an injunction that orders the infringer to stop the infringement.

Impact and Cost of a Preliminary Injunction

Among the most feared remedies in patent litigation is the preliminary injunction. This is an action or order from a judge that is issued at the beginning stages of the litigation. The argument that supports this is the assertion that the patent holder is suffering irreparable harm that cannot be remedied with money damages.

An example of such an argument is that you are facing a powerful competitor and your company will be destroyed and unable to pay for litigation if the infringement is not halted immediately. You could also argue that the competitor is unlikely to be able to pay any significant damages so that financial recovery will be limited.

The judge needs to carefully consider whether the patent seems likely to prevail in the litigation, the impact on consumers of halting sales of the infringing product, and the inability of money damages to fully compensate the patent holder. If the judge grants the preliminary injunction, however, the court could demand that the patent owner post a large bond. The bond solves the problem of how to compensate the accused infringer for its losses due to the injunction if the patent is later invalidated during the course of the litigation. The requirement to post a bond may be eased if the litigating patent holder has limited resources, but the bond requirement often means that a preliminary injunction is out of reach financially for the very companies that most need such an injunction.

Permanent Injunctions

In many cases, once a patent holder wins the patent infringement lawsuit, there will be a permanent injunction stopping the competitor from engaging in any further infringement. However, the courts will more often prefer to order the competitor to compensate the patent holder for its financial damages, but force the patent holder to issue a license, allowing the competitor to continue making the patented product as long as a royalty is paid.

Permanent injunctions are most likely to be allowed in situations in which a patent holder has kept the business solely to itself with no licenses in the past and in which the patent holder seems perfectly capable of providing the product to customers. Another factor is where a particular patented product will open up access to valuable customers and allow the patent holder to take advantage of this to gain sales of unpatented products. The inability to sell unpatented products is not compensated by patent damages and can only be remedied by an injunction that sustains the patent holder's exclusive position in the marketplace.

Licensing as a Result of a Win

In some cases, the money damages are calculated to cover past and future profits or royalties. In those cases, the license that the victorious patent holder is forced to give to the infringer is a fully paid up, royalty-free license.

If the competitor has to pay royalties in the future, this means that the patent holder must develop an ongoing business relationship with the defeated infringer. This is often seen as unattractive to both parties. By closing out the litigation with a comprehensive payment and a royalty-free license, the two parties have finished with each other and can go their separate ways.

Suing Government Infringers

Although you can sue private companies and individuals for infringement, governments are generally protected by sovereign immunity - a person cannot always sue the government.

The United States allows itself to be sued for reasonable patent royalties by filing in the US Court of Claims. The States of the US have been completely immune because you have to file suit in Federal Court and States have immunity against being sued for damages in Federal Courts. I have been the only person to succeed in suing a State for patent infringement - my company NeuroGrafix brought an "Inverse Condemnation" action against the University of California which led them to agree to face clams of patent infringement in Federal Court. This action was related to the laws of eminent domain where a State takes your property without compensation.

CHAPTER 19

The Right Attorney for Infringement Litigation

In This Chapter

➤ Considerations when choosing a patent litigation firm

➤ Different types of payment arrangements for law firms

➤ Motivation and teamwork during a patent litigation

The best way to win a patent infringement litigation is to have a great patent, but a close second is to have a great team of patent attorneys.

In patent litigation, you are unlikely to work with just a single attorney. This is because of the large amount of work that can be required in litigating a patent case. To succeed, you need a team leader who is an excellent litigator as well as a knowledgeable patent attorney. A litigator is someone who can stand up in front of a judge or a jury and convince them to see things from his point of view as opposed to the alternate reality being asserted by opposing counsel.

In addition to its verbal skills in the courtroom, the team has to be able to generate compelling written pleadings. These have to be extensively referenced as to previous court cases and rules that provide relevant precedents and guidelines. The team needs to understand your technology thoroughly and be passionate about winning, and this has to come through in every motion and in every opposition filing.

Choosing an Attorney

There are a variety of law firms, each with its own pluses and minuses, that have expertise in patent infringement litigation. Your choice of what type of firm to retain should be based on the features of the anticipated litigation.

Large vs Small Law Firms

The first step in choosing a lead patent attorney is to find someone with some experience in litigation in an area of technology close to your own. You also need to rely on personal referrals and any evidence you can locate about the successfulness of a given attorney.

In theory, at least, a large law firm with hundreds or even thousands of attorneys can rely on individuals within the firm who have expertise in particular issues that can arise during litigation. A large firm is also relatively immune to the financial risk of litigation.

If everything was looking promising for a big win but you were behind in paying your attorney's bills by $500,000, consider the impact on a small firm versus a large firm. That amount of unpaid money could drive your small law firm bankrupt. Even if a huge win is just around the corner, the amount of unpaid money could make further progress impossible.

A large firm with 3,000 attorneys, each collecting on average $1 million per year, has a $3 billion per year turnover. The large firm won't be happy that you are behind in your payments, but if prospects look good, your being in arrears won't have a significant financial impact on the firm.

The flip side of this financial equation is that the amount of billed legal expenses from your case could be so small in the firm's view, that your work gets low priority and little attention. The firm may be dealing with large corporate clients that are paying tens of millions of dollars per year. The amount of money you are paying may seem large from your perspective, but may be small from the point of a major multinational law firm.

Another important issue concerns the problem of conflicts. If you are suing a large technology firm, it is possible or even probable that the large law firm you're interested in hiring has done work for the very company that you plan to sue. If that is the situation, the firm will not be able to take your case.

If you have several major technology firms on your potential defendant list, you may be unable to find a firm that has never represented at least one of your potential targets.

A small newer firm, on the other hand, may have no conflicts at all, even though attorneys at the firm worked previously at other firms with conflicts.

This is much less of an issue if your infringer is a smaller, more specialized business.

Communicating With a New Set of Attorneys

Some technologies are fairly straightforward and can be readily understood by an attorney. But in those cases that involve a large, complex patent in a complex area of technology, it may take months before your law firm is fully up to speed with all aspects of your technology.

Patents Expert Pointer

It's often helpful to set aside a considerable amount of time with your attorneys to explain the technology and how it is expressed in your patent. They will also need to understand the prior art in your field as well as the structure and finances of the infringing entities.

A great deal of time is often required for an attorney to fully understand the patent itself. Claim language is complex, and it may take considerable time and effort for your own team to fully understand all of the phrases and references used. The patent needs to be read and reread many times by your legal team, and the file history that came out of the process of examination and grant needs to be fully understood as well.

General Firms vs Boutiques

Patent litigation is often just one capability in a large law firm. You could have a firm with a thousand attorneys, but just fifty who specialize in patent litigation. There are also large intellectual property firms in which there are, for instance, two hundred attorneys, all of whom specialize in patent litigation. At the other extreme are small firms with a single attorney who specializes in patent litigation.

A large law firm with offices in many cities provides an advantage when you have plans that may include litigation against a number of different defendants in different states or countries. If your law firm has only one office in the city where you reside, then you may need to retain local counsel with some other firm if you need to file an action or respond to a declaratory judgment in a different state.

Larger firms may have lead attorneys who charge high rates, but they may also have a number of junior attorneys with low rates who handle a number of time-intensive lower-level tasks (managing discovery documents, doing legal research, organizing complex documents for filing).

A smaller firm may view you as a more important client because it may have fewer high-dollar superstar clients, and the attorneys may improve their reputation and build their firm's fame and prominence with a successful outcome in your case.

Even among patent-oriented firms, some include a variety of intellectual property capabilities. Some represent trademark and copyright actions; others have a group of attorneys who have passed the patent bar exam who specialize in drafting, filing, and prosecuting patent applications. Out of all of this, only a few of their attorneys may actually specialize in patent litigation. In a true boutique firm, all the attorneys do is patent litigation.

Payment Agreements between Patent Owners and Litigators

Once you have picked out who you would like to approach to conduct a patent litigation, you may need to consider the various options for paying the costs of litigation. Although the average cost of a full patent litigation is around $3 million over three years, it includes a variety of expenses that may arise at different times. You could have low expenses one month and high expenses another month. Various legal fee agreements can be worked out that make it possible for solo inventors and small businesses to consider using patent litigation to protect their intellectual property.

Most attorneys and legal firms offer their clients a choice of payment options.

Contingency Agreements

In a pure contingency arrangement, you pay nothing at all to the law firm before or during the proceedings. Instead, the law firm gets a large percentage of any settlement or judgment (as high as 50 percent) once the case is won. This has obvious attractions from the point of view of the patent owner:

> ➤ There are no ongoing expenses

> ➤ Although a loss may present significant problems for the business, it will not result in a huge uncompensated legal bill

> ➤ It will seem to the patent owner that this arrangement provides a powerful motivation for the law firm to do the absolute best possible job it can do

For the law firm, this does provide an opportunity to collect considerably more than it would by fees alone if the patent covers a product of high value with large expectations of damages.

Obviously, any law firm considering taking on a case on a contingency basis must have a high level of confidence in the likely success of the patent litigation. Even after the attorneys

Patent Vocab

Contingency is a fee arrangement in which a law firm, in this case, is paid a percentage of any collections from a successful settlement or victory in the courtroom. Without a contingency, the attorneys get paid the same, win or lose.

have done millions of dollars of work, if they don't win, they get nothing. If the situation was that the only way they would lose is if they did a poor job as attorneys, then this might be fair. However, the loss may take place despite excellent and passionate representation.

Hourly Agreements

The opposite standard method of paying for legal services is to have a monthly bill representing payment by the hour. Attorneys could charge anywhere from $200 per hour up to $1,000 per hour. Because of this, a monthly bill of up to or over $100,000 should not be a surprise. If you work out the numbers, the $3 million over three years amounts to $83,000 per month.

This kind of monthly bill may overwhelm a small firm even if it looks likely to win. The attorney representing the firm may be enthusiastic, but other partners in the law firm may be anxious about potential unpaid bills and may feel that your attorney is overly optimistic.

Mixed-Fee Agreements

One way to help handle the expense is to do a more complex mixed-fee agreement. There might be a fixed monthly amount due to the law firm that is applied to the balance. The fees may be discounted (charged at a lower hourly rate) in exchange for some contingency amount—for instance, 10 percent or 20 percent. There may also be an agreement in advance that any unpaid amounts would be paid off over time (for instance, thirty-six months after the close of litigation).

Project-Based Agreements

During the economic downturn starting in 2007, many law firms came under tremendous financial pressure as the volume of business litigation dropped. In some cases, this allowed clients to get law firms to agree to project-based fees. Instead of an hourly rate, the law firm would agree to various payments for accomplishing particular segments of the case in satisfactory fashion.

This kind of arrangement forces the law firm to be economical with its lawyers' time so that the fixed payment is sufficient to pay the lawyers for their time. At the same time, it maintains pressure on the attorneys to do excellent work because the fee agreement has various cutoff points that the client (the patent owner) could take advantage of to terminate representation and switch to another firm.

Nonpracticing Entities (NPEs)

A law firm may also reach out to investors and NPEs (nonpracticing entities—firms that invest in patent lawsuits but don't manufacture or sell the patented product [see Chapter 18]). In this situation, the NPE invests in the litigation in the law firm. This separates NPEs from involvement in patent owners' businesses and helps assure that they won't be drawn into any expenses or counterclaims that the patent owners may face.

Keeping an Eye on Your Attorney

In theory, of course, every attorney represents his clients with maximum enthusiasm and energy. However, your attorney may be handling multiple cases that all make great demands on his time. There may be a natural tendency for an attorney to devote more attention to an effort that is most likely to bring in the most income—and that may not be your case.

In large law firms, this may result in some of the work being shifted to junior attorneys fresh out of law school. In some cases, law firms have even hired overseas attorneys (think

Patents Expert Pointer

Even after you have selected and retained a patent law firm for an infringement litigation, the patent expert knows that the only way to get the best result is if the lead attorney really understands the patent and is honestly enthusiastic about its potential success. Just before the litigation is filed, you should take the time to be sure that the buildup to filing has confirmed the wisdom of your initial choice of patent firm. It is much easier to change firms before filing than after.

Mumbai) to do labor-intensive research and motion-drafting work. The senior attorney then reviews the work to make sure everything seems to be in order.

The problem is that the lead counsel for the other side may be extremely well paid and can be devoting tremendous skill and expertise to attacking your position, while your firm produces half-hearted or inadequately prepared briefs.

It is difficult for the inventor to know when this is happening. Of course, when you win, you feel your attorneys are heroes. When you lose, you may suspect they could have done better. The best you can do is to learn about the patent litigation process and try to be both vigilant and helpful. The more you know about the process, the better job you can do at managing and optimizing your legal representation.

If you don't meet your financial obligations, the law firm can back out of representation. In fact, a law firm can also back out because it is being pressured by other clients. This is highly

Inventor Beware

It is important for a company to understand how the expenses will flow in during a patent litigation. If you can't reasonably keep up with the expenses you have agreed to pay, you may get abandoned by your law firm with no means of continuing the litigation.

unethical for the lawyers, but large law firms can face complex pressures from various clients who may be far more important to the firm than you are. You are more likely to see pressure to discontinue representation for poor payment when you are represented by a smaller firm, and more likely to see this in response to external pressures on the firm if you have retained a large firm with many potential conflicts.

If things are going well financially (you are keeping up with your bills) and there are no external pressures on your firm to back out of representation, it may still be the case that you become dissatisfied with the performance of your attorneys.

If large, complex documents are being drafted at the very last moment, if the attorneys seem to be unprepared when questioned by the judge, if they continue to make mistakes that show they are not taking the time to understand the patent fully or to fully review and respond to a brief filed by the other side, then you may want to think about terminating representation and moving on to a different firm.

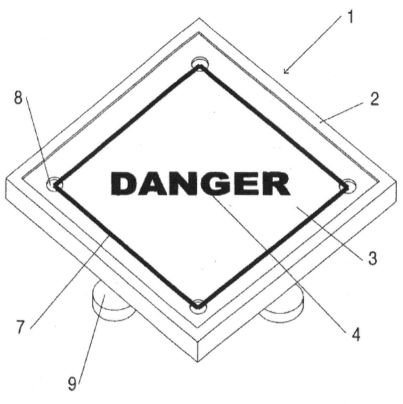

FIG. 1

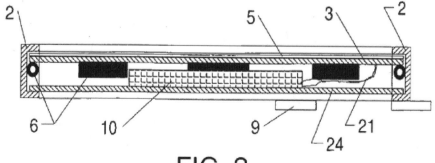

FIG. 2

 Defending Validity and Constructing Claims

---

In This Chapter

➤ Facing a challenge to the validity of your patent

➤ Why a Markman hearing is critical to winning or losing

➤ The role of the federal circuit in hearing appeals

If you ever do file a lawsuit against a competitor alleging patent infringement, don't expect an apology and an inquiry into how large a check he should write out to you. In fact, after completely denying he has done anything wrong, the very next thing the competitor will do is sue you and try to invalidate your patent.

Facing a Validity Challenge

There are a number of grounds under which an opponent will deny having any liability. These come out of a list of standard defenses. It would be malpractice by the attorneys representing your opponent if they did not deny everything and file a suit against you. This is more or less required, so try not to be offended, outraged, or depressed.

Getting Sued

An infringer's first line of attack may be that you don't have the right or standing to file the lawsuit because of a fault or problem in the exclusive license or original filing details of your patent application. The nastiest kind of action an opponent can take is to allege that you were only able to obtain the patent in the first place by engaging in fraud.

Inventor Beware

Once you have decided to file a patent litigation, you have also decided to risk having your patent invalidated. In some cases, the potential impact of getting invalidated may outweigh any potential benefits that could be gained from infringement litigation.

The infringer will assert that you haven't properly disclosed enough in the patent text to allow anyone else perform the invention. He will assert that you didn't know what you were doing, and anyone performing the steps described in your patent will not get a working product. He will also allege that the language of your claims (written by your patent attorney) have various legal defects and don't match your patent description.

Remember, if the infringer's attorneys lose, their client may face huge damages, so they will spare no effort in trying to stop you. You had better be prepared for all this or you should not consider undertaking an infringement action.

Federal District Court

When your patent is being examined by the patent office, you are dealing with people who want you to get a good-quality patent granted. They are more or less on your side. They are experts in patents; patents are what they love in life.

Your examiner will know a great deal about your area of technology, prior inventions in your field, and the general pace and quality of the progress of innovation. He will almost certainly share your belief that inventions are the key to advancing the human condition, and inventors are heroes to be respected and encouraged. It's true that examiners are exacting and don't want bad patents to issue—that is their mission. However, they are still fundamentally boosters of the invention process.

The situation in federal district court is different. The judge who is assigned your case may dislike patents, dislike patent litigation, and harbor great suspicions about the motives of inventors and the attorneys who represent them.

Suppose that this was a litigation about personal injury in a car accident. Your doctor was supportive and your attorney assured you that you deserve a large settlement from the insurance company of the driver who caused the accident. However, the judge may have seen scams and misrepresentations and may feel you are just one more person trying to take advantage of the system in order to cash in. This kind of attitude can also affect a judge's view of patent litigation.

Overall, the vast majority of federal district court judges are fair, impartial, knowledgeable, involved, and committed to justice. However, a judge understands that at the outset of the case, it is not clear whether justice will be done by invalidating your patent or by finding against your opponent. The whole reason that we have a legal system is that there are two sides to a dispute. Both sides may have strong elements of justice in support of their position, and the difference between victory for one side or the other may reflect a narrow margin of advantage in favor of one of the parties.

If you have spent a year responding to motions and battling over procedures that have little to do with the substance of your invention, then you may become angry and frustrated by the process. These aspects of a patent litigation exist because they are important. They have grown out of centuries of experience in dealing with disputes of all types and with patent disputes in particular.

Once it does become time to deal with the substance of the actual technological dispute, you may again be surprised by the formal and legalistic way that the process is carried out. It's a legal system, and you are trying to enforce legal rights, so there is not really any option to do this any other way.

Claim Construction in a Patent Litigation

A major transformation of the patent litigation process, after hundreds of years in which it was handled more or less just like any other property or contract dispute, was the development of a process called the Markman hearing. This took place following a Supreme Court decision in 1996. The purpose of the hearing is to carry out a process called claim construction, which results in an agreement among the plaintiff, defendant, and judge about exactly what each word or term in the most important parts of the claims actually means.

The Critical Role of the Markman Hearing

The case that led to the creation of the Markman hearing was *Markman v. Westview Instruments Inc.* In that case, Markman had a patent that related to a method of monitoring the progression of clothing through the dry cleaning process. Westview Instruments was using a system that tracked charges for steps of the dry cleaning process. Markman alleged that the system used by Westview infringed on his patent. When the case finally got to the jury, the jurors decided in favor of Markman. However, the judge overruled the jury and decided in favor of Westview.

Patent Vocab

A Markman hearing is a legal proceeding that allows a judge to make decisions about the meaning of important words and phrases in a patent. Very often, a patent litigation is won or lost based on the result of the Markman hearing.

The issue was the meaning of the word "inventory" in the patent claim. If the word was used to include both financial issues and actual articles of clothing, then the Westview system didn't infringe since it didn't track actual articles of clothing.

The jury heard expert witness testimony stating that the word "inventory" in this setting referred only to the financial tracking and accepted this view in finding in favor of Markman. The judge felt that the jury had gotten it wrong, and this one issue proved to be the deciding factor in the case.

This dispute went to the Supreme Court as a broader question: should the determination of the meaning of words be made by a judge or by a jury?

Historically, the problem of construction of terms and delineations of boundaries came up all the time in contract and property disputes. The meaning of words used in contracts and property boundary disputes were always decided by juries. This is because in our legal system, a judge is supposed to make decisions on matters of law, and a jury is supposed to make decisions on matters of fact.

The whole basis of a jury system is that the two sides will have different descriptions and views of the central facts of the case. The twelve jurors are supposed to tell the court how an ordinary, reasonable person would understand the words.

The ordinary, reasonable person standard seemed to apply to ordinary situations of contracts and property. The Supreme Court decided that words in a patent claim are specialized and reflect small individual rules of law. As such, the Supreme Court decided that the construction (interpretation of meaning) of the words in a patent claim are a matter of law to be decided by a judge and not a matter of fact to be determined by a jury.

In a jury trial, all the motions and all the twists and turns of discovery, and the depositions go first before the case is finally ready to go to the jury. There may be three years of buildup and one week in front of a jury. Under the old system, neither side really knew the true strength of its position until the final moments of the litigation.

The institution of the Markman hearing was intended to change all that. Now, the two sides choose a list of important terms, argue about the meanings before the judge at a hearing, and receive a decision on the meanings early in the case.

Ideally, the order generated in the Markman hearing would bring about the termination or settlement of the case without the full time and expense of a jury trial in many cases. Both sides would be able to see exactly what the patent covered and whether or not infringement had taken place.

How the Judge Decides the Meanings

In deciding the meaning of the terms, the judge hears evidence that is ranked in order of decisiveness. The highest level of impact goes to any specific statement made in the text of the patent for which the inventor provides a definition. This process is called being your own lexicographer. Even if the standard or common meaning of a word is different from your definition, it is the definition given in the patent description text that is definitive.

Aside from specific definitions, the judge looks at any comments or information that is in the intrinsic record. This means anything said anywhere in the patent or anywhere in the exchange of communications between the patent examiner and the inventor during the course of the patent examination process—as well as any appendices submitted with the patent application originally.

If nothing within the four corners of the patent and the file history from the patent exam settles the question, the judge will probably look next to technical dictionaries that were available at the time the patent was filed. Here, the two sides may present different dictionaries that show different or competing meanings.

Patents Expert Pointer

Well before the Markman hearing, the patent expert begins accumulating technical dictionaries that were available at the time the patent was filed. If, despite all precautions, the other side identifies technical terms used in your patent that may be ambiguous, these dictionaries can be your best hope of fighting back.

Dictionaries are extrinsic evidence but are given special regard in this process. Beyond this, the judge may look to testimony in depositions of experts who will give their opinion on what a given word would have meant or how it would have been interpreted at the time the patent was filed.

In some cases, the judge may actually decide that there is no way to determine what a critical term actually means. If that happens, the claim may be declared invalid because it is indefinite.

The Court of Appeals of the Federal Circuit

The expectation of using the Markman hearing for claim construction to encourage early settlement has proven true to some extent, but overall there has been little real impact on the tendency toward early settlement. The problem is that if the two sides don't agree on the interpretation of terms made by the judge, there is likely to be an appeal. However, an appeal does not take place until the case has gone all the way through to completion of the trial in front of the jury. Only after the jury verdict can the disappointed side file an appeal with the Court of Appeals of the Federal Circuit (CAFC).

Some members of Congress have suggested that litigants should be allowed to appeal to the CAFC immediately after the Markman hearing without waiting for the rest of the trial to go forward. This would be an interlocutory appeal. However, this was not included in the AIA patent reform of 2011.

Once the CAFC has reviewed the case, it may issue a critically different interpretation of the terms in the claims. In many courts of appeal, the judges involved give deference to the decision made by the federal district court judge (the lower court). This means that it needs powerful convincing to decide that the federal district court judge got it wrong.

In patent cases, however, the CAFC gives no deference to the decision of the district court judge. It looks at the facts and arguments provided and makes up its mind de novo—a fresh decision with no special regard for what the district court judge decided. This is because the CAFC is a specialist patent court and has more expertise in patent matters.

The result has been that the CAFC is more likely to change or reverse critical definitions made by the district court judge than it is to confirm them. This has led to the humorous assertion that the best way to win in this area is to have the federal district court judge rule against you and then go on to appeal.

The Supreme Court's View on Claim Construction

In some cases, the decisions made by the CAFC are appealed to the Supreme Court. There is no obligation for the Supreme Court to agree to hear an appeal of a CAFC decision.

It is most likely to agree to hear a case if it believes there is an important general rule of law that can be resolved by considering the special facts of a given case. In general, the Supreme Court has refused to hear appeals of CAFC determinations on claim construction apparently on the grounds that these cases have not tended to pose any broad or general question of law.

There have been challenges to the CAFC position that it does not need to give any deference to claim construction decisions made by federal district court judges. The Supreme Court has refused to hear three of these challenges that have arisen since the CAFC laid out this position in 1998. For these reasons, the CAFC position that it should make a fresh review of claim construction decisions and that its review is final seems to be well-settled law at this time.

T. A. EDISON.
ELECTRIC LAMP AND HOLDER FOR THE SAME.

No. 265,311.

Patented Oct. 3, 1882.

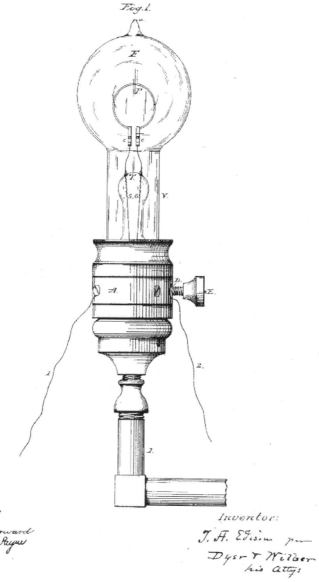

Fig. 1.

Attest

F. W. Howard
James W. Payne

Inventor:

T. A. Edison per
Dyer & Wilber
his attys

N. PETERS, Photo-Lithographer, Washington, D. C.

 Famous Wins and Losses in Patent Litigation

Overall, patent litigations are fact intensive. This means that it is not always easy to take away lessons from one case and apply them to another. Nonetheless, it is interesting to look through some of the most dramatic cases that have been litigated to look for general lessons.

The Biggest Wins and How They Happened

It's not surprising that some of the biggest wins in patent litigation were in cases that involved some of the largest companies.

Polaroid vs Kodak

One of the largest and most complex litigations in the history of the patent courts involved a battle between Polaroid and Kodak over instant cameras. The two companies had worked together in the 1960s, but in 1976, Kodak introduced instant cameras that were eventually found to infringe Polaroid's patents. Polaroid sought $12 billion in damages.

The case did not get resolved until 1991—fifteen years later. The damages awarded were about $900 million paid to Polaroid by Kodak. Kodak abandoned the instant camera market in 1986. The actual trial went on for ninety-six days over eight months. Much of the litigation dealt with complex issues about the determination of the scale of the damages.

Polaroid eventually went bankrupt in 2001, largely because the arrival of digital cameras destroyed the market for instant film cameras. Kodak went bankrupt in 2012. It had made the transition to digital cameras more effectively than Polaroid, but the loss of its film business to new technology and the effect of camera phones, such as the iPhone, on digital camera sales eventually overwhelmed it.

Some have argued that both companies might have done better in adapting to the advance of technology if they had not been so consumed by litigation; however, there is no way to prove this. Both companies were large enough that the litigation did not clearly affect their ability to do research and develop technologically.

Michelson vs Medtronic

Gary Michelson is an orthopedic spine surgeon who developed a series of technologies and patents used in spinal surgery. For years he worked with Sofamor Danek (later purchased by Medtronic) in developing and marketing the inventions. He came into conflict with Medtronic in a series of marketing and patent lawsuits and countersuits. The $1.3 billion settlement came in 2005 after Michelson spent $62 million in litigation expenses in litigation expenses during 11 years of disputes and a final four year litigation.

The eventual settlement has been portrayed as an acquisition by Medtronic rather than as a settlement of patent damages. The result has Michelson working closely with Medtronic, licensing over two hundred of his patents to them, and promising to license his future inventions to them over the next fifteen years on a royalty-free basis.

When you consider that Medtronic had been paying $40 million per year in royalties, you can see that future royalties would have been $600 million. The numbers make sense in light of the fact that the market for spinal implants is about $3 billion per year, and Medtronic is the leading firm in that market. Medtronic had cash reserves of $4.5 billion at the time it agreed to the settlement.

Damadian vs General Electric

Raymond Damadian was famously denied participation in the Nobel Prize for the invention of MRI (Magnetic Resonance Imaging). However, his $100 million win in patent litigation with GE did help secure his position in history as the inventor of MRI. This was a hard-fought technical battle because Damadian's scanner lacked many of the components and functions of the first commercial MRI machines.

The case involved a second patent coming from his firm—FONAR—that offered a key software function that GE had advertised as well. The case went to appeal in the CAFC over the issue of whether the patent could be valid if it did not provide the full source code to explain how its software worked.

The CAFC overruled the federal district court on this. GE appealed to the Supreme Court, but it declined to hear the case. This case also helped establish that full source code is not needed to demonstrate the best mode for carrying out a particular software function.

TiVo vs Dish Network/EchoStar

Almost everyone now uses a DVR (digital video recorder). Many will remember when this clever new technology first came out and everyone was buying TiVo units. Since that time, TiVo has seemed to fade away as DVRs were offered widely by a variety of competitors. However, between 2004 and 2011, patent infringement litigation progressed. TiVo won a patent infringement verdict against EchoStar and the Dish Network. The judge issued a permanent injunction halting Dish Network/EchoStar from selling DVRs.

This was appealed to the CAFC, which distinguished hardware aspects of the DVR from software aspects, but supported the injunction since it felt that the infringement of the software claims were solid.

Dish Network/EchoStar began selling a modified system, but TiVo won on a claim that Dish Network/EchoStar was acting in contempt of court by ignoring the injunction.

In May of 2011, the two companies settled the matter with a $500 million payment to TiVo and a license under the patent to Dish Network and EchoStar.

Dramatic Losses and Reversals

One repeating theme is that large wins in front of a district court can be reversed either by appeal or by action of the presiding district court judge if he does not agree with the decision of the jury.

Inventor Beware

If the federal district court judge cannot be convinced to respect and believe in your patent, the other components of the litigation will not be able to overcome this. An excellent, knowledgeable, and convincing presentation, including PowerPoint slides, at the first hearing in front of the judge is critical. This is when the judge makes his first impression, and it may be nearly impossible to change his views later in the case.

Alcatel-Lucent vs Microsoft

This was a complex litigation involving multiple patents from the Bell Labs/AT&T/Lucent portfolio of inventions. The patents included MP3 audio technology, MPEG video, and speech recognition technology that were asserted against Gateway and Dell as intermediate customers using and selling the Microsoft Windows operating system. Microsoft asserted patents against Alcatel-Lucent as well.

One part of the case led to a $1.5 billion jury verdict against Microsoft, but a judge later threw out the entire verdict and damage amount. The two companies (Alcatel as current owner of most of the Lucent technologies) and Microsoft have settled much of the dispute but continue to fight some parts of this multiple litigation patent dispute.

Patents Expert Pointer

When large amounts of money are involved, a win can turn into a loss as various rounds of appeals progress. Often the best way to win is to reach a settlement agreement that is overall favorable to your position. You don't get all that you believe you are entitled to, and your opponent may get a better deal than you'd prefer, but a settlement avoids the possibility of expense and uncertainty stretching out for decades into the future.

Medtronic v BrainLab

In this twelve-year litigation resolved in 2010, Medtronic alleged that BrainLab had infringed on its patents for computer-guided surgery. BrainLab had a good result in the Markman hearing when the judge determined that there was a critical difference in meaning that distinguished between the way the two navigation systems worked.

BrainLab brought a motion for summary judgment, asking the judge to end the trial based on the significance of the claim construction finding. The judge denied the motion, letting the case proceed.

At trial, the jury determined that BrainLab was guilty of infringement. The judge then threw out the jury verdict because it did not seem to accommodate the full implication of the claim construction decision.

BrainLab then alleged misconduct on the part of Medtronic's attorneys, saying that they had misrepresented the meaning of the claims in statements to the jury, ignoring the judge's claim construction order. The judge agreed and awarded attorneys' fees to be paid to BrainLab and assessed a large fine against Medtronic's attorneys (McDermott & Will).

Medtronic then appealed the punitive fees and fines to the CAFC. The CAFC overturned these actions and voided the award of attorneys' fees to BrainLab and wiped out the fines against McDermott & Will.

Litton vs Honeywell Reversal

This case dealt with infringement by Honeywell of components of an aviation guidance system. The case led to a $1.2 billion verdict against Honeywell, which would have been the largest verdict ever in a patent litigation. However, the district court judge (Mariana Pfaelzer) overturned the jury verdict and determined that the patent was invalid.

Litton appealed to the CAFC and won a decision restoring the verdict. However, appeals continued to reverberate. There were multiple issues addressed in the district court and the CAFC, including inequitable conduct, patent validity, claim construction, and willful (intentional) infringement.

A Supreme Court decision on a different case during this time substantially altered some of the critical legal reasoning used. In the end, after eleven years of litigation and appeals, Honeywell settled the case, agreeing to pay $440 million to Northrop Grumman, which had acquired Litton.

Complex Multielement Conflicts and Patents

For most inventions, there is a single idea and a single product. However, modern products can be astonishingly complex. Many inventions can contribute to a new machine. The patent courts have been struggling with this problem of interdependence of technologies, and no clear solution is in sight.

The Matter of Android and the iPhone

Millions of users will agree that the iPhone and iPad were game changers and seem to have many unique and innovative features relative to previous phones. However, the Android phones with software from Google look very similar to the Apple iPhone. Steve Jobs was outraged and said this reflected "slavish copying" of his groundbreaking innovations. This has led to an array of very large complex patent litigations.

Oracle has lost the first round in its suit against Google for $3 billion over its use of Java software in Android. Apple won a $1 billion victory in its suit against Samsung. Samsung lost in US Court, but won in Korean Court in its suit against Apple for its tablet. Samsung is suing Apple over the use of some basic patents in wireless technology. Google purchased

Motorola for $12.5 billion so that it could start litigating against Apple with Motorola's patents. Many of the patents in the Motorola portfolio relate to standards that are used

Patent Vocab

Standards patent refers to a key innovation that becomes the basis for a wide array of subsequent technology. If all of the competing companies in an industry must have access to a right to use a particular standards patent, the courts have tended to force the owner of that patent to allow licensing to all the competitors on reasonable terms.

across the industry in order to make various kinds of sharing of networks possible. Apple is complaining that Google is now overcharging for these "standards" patents.

Overview and Perspective

In the complex thicket of interdependent patents in the iPhone-Android battle, it appears likely that these various companies will end up with settlements. However, it seems as if there should be some sort of protection for truly novel inventions that has greater force than the patents for underlying and marginally unique components. There are obviously many different factors at work here:

➤ The vast resources to carry out the litigation among huge competitors

➤ The thousands of patents and licenses that go into a complex multifunction device such as the iPhone

➤ Standards that should be available for licensing by all

➤ Unique products that reinvent how we interact with each other and with the world at large

Nonetheless, despite the massive grouping of patents and technologies in the iPhone-Android conflict, there is still clearly an important place in the world for the individual inventor with vision and insight.

The patent system has helped accelerate innovation and rewarded inventors. The system is not at all perfect and has many inequities. Nonetheless, a good knowledge of the patent system should allow the modern inventor to progress from the flash of insight to the opportunity to affect and advance the human condition.

APPENDIX

The appendix for this book is too lengthy to be included in this volume and has been posted on our Internet site www.smartguidepublications.com.

You will find it on the Appendix tab under "Patents Book Appendix" as a PDF file which can be downloaded and printed.

INDEX

ABOUT THE AUTHOR

Aaron Filler, MD, PhD is an internationally prominent inventor and technology entrepreneur who has served as CEO or Chief Scientific Officer of several successful technology companies he founded—including venture capital- and business operation-funded businesses in the US and UK. He has served on the faculty at UCLA and the University of London and has been a Medical Director at Cedars Sinai Medical Center in Los Angeles. He has a PhD from Harvard University. He also helped manage a large-scale research effort based in companies he founded in London and near Cambridge University in the UK.

He is the sole or lead inventor on numerous granted US and international patents for significant advances in chemistry, pharmaceuticals, medical devices, and medical imaging. Dr. Filler also has many years of experience in navigating his patents through successful patent infringement litigation against a wide variety of patent infringers. His legal breakthrough piercing sovereign immunity that forced the University of California to appear in Federal Court to answer charges of patent infringement was widely reported in the legal news media.

In addition to extensive academic publications, he is the author of an Oxford University Press book on spine and nerve pain and a thousand-page book on Apple computers (Datamost, 1984). He has designed and written and currently maintains several large websites.

Public interest in his medical imaging and pain treatment inventions has led to coverage of his work in The New York Times, The Los Angeles Times, The Economist, and The London Times. He has been interviewed on CNN and in a segment on ABC World News.